MASONRY AND CONCRETE WORK

A POPULAR SCIENCE BOOK

MASONRY AND CONCRETE WORK

MAX ALTH

Drawings by

CARL J. DE GROOTE

POPULAR SCIENCE

HARPER & ROW

NEW YORK, EVANSTON, SAN FRANCISCO, LONDON

ACKNOWLEDGMENT

The author thanks Henry Gross, his editor at Popular Science Books, for his help and patience in the preparation of this book.

Library of Congress Catalog Card Number: 77-6555
ISBN: 0-06-010143-1

Manufactured in the United States of America

To Char,
Misch,
Syme,
Arabella
and
Mendel

CONTENTS

MASONRY AND CONCRETE WORK

WORKING WITH STONE

No one remembers the first man to place one stone atop another to block the entrance to his cave. Perhaps he was decried as a dangerous innovator. Nevertheless dry-stone masonry — stone construction without the use of mortar as a permanent "glue" to hold the stones together — has a long and honorable history. The pyramids were constructed dry, as were many cathedrals in Europe. Notre Dame is just one example of what men with skill and even more patience can do with stone alone.

When you use mortar with stone, construction is greatly simplified. The need for accurate cutting and fitting is sharply reduced. That is why few, if any, buildings are presently constructed of stone alone.

This portion of the book is devoted to the rudiments of stone masonry, dry and with mortar. Once you learn the basics of working with stone, you will be surprised at the variety of permanent improvements to your home and grounds you will be able to construct with only a few tools. Not only will these improvements increase the beauty of your property, they will add considerably to its value. I do not believe it would be far from wrong to estimate that the value of your property will increase about $4 for every dollar's worth of stone you install. And if you use fieldstone, which is literally stones picked up in the field, the dollar value of your improvements — at this time — could be reasonably computed at $150 to $200 for every full day's work you put in. Moreover all this will be in addition to the pleasure you will derive from working with stone, constructing things that will be as permanent as anything man ever constructed.

TOOLS FOR WORKING WITH STONE

Mason's Chisel

Small Sledgehammer

FLAGGED WALKS

One of the simplest, easiest, and least expensive ways to improve the appearance of a yard or a garden is to add a walk. And the easiest, least expensive type to build is a flagged walk — which is so easy you can lay 50 feet of walk in an afternoon, without experience and without any special tools. A flagged walk is one that is paved with flagging, which is any flat-sided material that can be laid down and stepped on. The following paving materials are most often used for flagged walks:

Flagstone   Bluestone   Brick   Fieldstone
Slate   Concrete block   Belgian block

Flagstone. This is a kind of sandstone that occurs naturally in layers. These vary in thickness from about ½ inch to several inches. The stones the yardman calls 1-inch will vary from ⅝ to 1½ inches in thickness. The ones you want are called 2-inch stones and vary in thickness from 1½ to 2½ inches. Only the 2-inch stones are strong enough to withstand cracking when used as described here. (You can also use 1-inch stones, as we'll explain later, but this requires much more preparatory work.) Flagstone can be cut with a power saw and a diamond or an abrasive blade, or by hand as described on pages 10 and 11.

Slate. Like flagstone, slate is a form of sandstone. It is closer grained than flagstone and much harder. Three types are commonly used for general construction: thin slate, $\frac{1}{8}$ inch thick; slate $\frac{3}{8}$ inch and thicker that is fairly smooth and uniform in thickness; and textured slate, a single piece of which may vary in thickness from $\frac{3}{8}$ to 3 inches.

Slate can be obtained in various colors—gray, green, dark blue, purple, and red. Gray is the least expensive. Rectangular pieces of slate, with cut and polished edges, are used for sills and occasionally for steps. Fairly flat, perfectly cut slate is too expensive for paving large areas. Therefore the textured stone is most often used for this purpose. Unfortunately, these pieces are usually irregular in outline, with one or more edges uneven and broken.

Slate can be cut like flagstone with a power saw and a diamond or an abrasive blade. The alternative is to use a nibbler, which is a sort of crushing tool that breaks off a little of the stone at a time. This action can be duplicated by holding the edge of the slate at an angle firmly against a stone or concrete surface and crushing that edge by repeatedly hammering it. Each stroke of the hammer breaks off a half inch or so. The method wastes both stone and time and will not produce a smooth, straight edge.

Bluestone. Still another form of sandstone with a grain somewhat between that of slate and flagstone, bluestone isn't carried by many yards these days. However, with all the urban renewal going on, you may be able to secure this stone for the asking. Bluestone can be hand-cut exactly like flagstone (see pages 10, 11).

Concrete block. This category includes standard, solid concrete block, the new white and colored block, and the block that you can make yourself by pouring concrete into a form.

The solid concrete building block most often used for paving is the 4-inch block. Its actual size is $3\frac{5}{8}$ by $7\frac{5}{8}$ by $15\frac{5}{8}$ inches. Garden block is available in a number of colors. Its overall dimensions are identical to the 4-inch building block, but it is only 1 or 2 inches thick.

Brick. Common brick (also called standard) is the least expensive of all the many types of brick currently manufactured. Used common brick is even cheaper and often has a texture somewhat like expensive, handmade brick. New brick is a soft red color. Used brick varies in color from pale yellow through deep red to black, often in the same brick. Common brick is $2\frac{1}{4}$ by $3\frac{3}{4}$ by 8 inches in size.

Difference between a flagstone edge that has been cut with a chisel (upper slab), and one that has been sawed.

Slate, unlike flagstone, cannot be cut with a chisel but must be broken with a small sledgehammer.

Precast concrete paving slabs create a handsome walk in front of this ranch-style house.

Belgian block. These are hand-cut pieces of gray granite that have been used for paving for hundreds of years. Generally they are 6 by 6 by 16 inches in size.

Fieldstone. The least expensive of all the flagging materials you can use is field-stone, which is literally stones you find lying in the field. If you are lucky you may find all the stones you need in your backyard. If not so lucky you may have to drive to a construction site where you can usually have as much stone as you wish for the asking.

CONSTRUCTING A FLAGSTONE WALK

The technique is simplicity itself. You purchase stones just as long as the width of the walk you are going to construct. The width can be anything you desire, but generally walks, other than those made with stepping stones, are at least 2 feet wide. Front walks are usually 3 to 5 feet wide. Sidewalks are usually 5 feet wide.

Flagstones are generally precut at the quarry or yard into squares and rectangles. A 2-foot-long flagstone is common enough, so let us suppose your walk is to be 2 feet wide. The width of each stone is unimportant. It can be any width that won't turn underfoot. If you wish, use stones of varied widths to make the walk more interesting and informal.

Start by cutting all the grass on your lawn as close to the ground as you can or by just cutting a swath through the grass 2½ feet wide along the route your path is going to follow. Carefully sweep this path clean of all debris.

Beginning at one end, carefully lay down the flags, one after another, with a small space between them. (Use a length of 1x2 board to help you space the flags evenly apart.) If you find, on coming to the end of the walk, that the last flag is either too wide or too narrow to fit the remaining space properly, you can readjust all the spaces between the stones, or ask the supplier to exchange the remaining stone for one that is the right size. When the last flag is in place to your satisfaction, spread loose soil between the flags and along their edges. Scatter grass seed, tamp firm and you are done.

This kind of pavement is classified as a flexible pavement and has a lot of advantages in addition to that of easy and economical construction.

For one thing, a flexible pavement cannot be damaged by frost. No matter how

Examples of how a walk can be made with flagstones of two different sizes. Left, a narrow stone follows a wide stone. Right, wide stones are laid across the path, narrow stones side by side. Board is used as a spacing guide.

high the earth may heave in response to cold, the earth and the flags on top will eventually settle down in the spring. And if the earth subsides and the stones sink a little, as they will when the grass underneath rots and the topsoil compacts, you can always lift the flags, toss a little soil or sand beneath them and let them lie there happily for another few years.

A simple variation is the stepping-stone walk. The construction procedure is exactly the same. You can use smaller flags in this case, but they should not be less than 1 foot wide or they will not long remain level lying on cropped grass and soft soil. And do not use flags thinner than 2 inches for the cracking reason already mentioned. Just lay the flags down in a line as close together or as far apart as you wish.

Alignment. In both the examples described the flags were laid directly on shorn grass, their alignment being judged by eye. If you wish for greater accuracy, using string and pegs stretch a straight line alongside the path you wish your flags to

Constructing a flagged walk of 2-inch-thick concrete block. First, stakes are driven into the ground at each end of the walkway. The distance between stakes is slightly greater than the length of the blocks. Then strings are attached and measurements checked. Blocks are laid in place using a board as a spacing guide.

To make a turn in a flagged walk, drive a stake into the ground to locate the center point of a "compass." Tie one end of a string to the stake and use it like a compass to describe an arc, driving stakes into the ground as you go. Then lay the blocks to form a curved path following the string guide.

follow. Lay the flags with their sides close to, but not touching, the line. If each flag presses even gently against the line, by the time you have laid a half dozen flags the line will be pushed to one side.

Turns. You can turn your walk by simply placing the flags in a curve or you can lay out a perfect curve using a peg and a string as a giant compass. Drive the peg into the earth and tie one end of the string to it. Then swing the end of the taut string in an arc around the peg. Drive more pegs into the ground as you walk to mark the curve the moving end of the string has described. To lay out a second curve parallel to the first, increase the length of the string by the width of the desired walk and repeat the operation.

It is easy to make turns with stepping stones. No one notices the changes in spacing between the small flags as they make the turn. However, when you work with flags more than 2 feet in length, the long triangular spaces between the flags forming the turn can become too obvious.

There are three solutions to this problem. One is to make the turn so gradual that the triangular spaces are not noticeable. Another is to bring a sketch of the walk to the masonry yard and have the flags cut to fit properly. The last is less expensive than the second and far more healthful. It consists of cutting the stones yourself.

Cutting flagstone. You can saw through flagstone with a high-powered saw and a diamond or an abrasive blade. However, diamond blades are expensive and abrasive blades wear out very quickly. The far more economical method of cutting flagstone is to use a stonemason's chisel about 3 inches wide and a small sledgehammmer.

Using a chalk line or some other means clearly mark the line of cut. Lay the stone on a 2x4 with the line centered over the length of the wood. Place the edge of the chisel on the line and rap the chisel lightly but repeatedly, sliding it along the line as you do so. After you have gone along the line a few times, you will have cut a groove. Turn the stone over and repeat the same hammering operation on the underside. When you have grooved the second side, slide the stone across the supporting board until the line of cut is about an inch beyond the edge of the wood. Tap the end of the stone lightly with the hammer until the stone snaps at the line. With a little practice you will be able to snap-cut a 1-foot-wide stone in about twenty minutes. Inside curves cannot be made this way. Outside curves are made by snap-cutting segments, one at a time, until an arc is formed.

One point before going on. You can make your walk of irregular flagstones and thus save money. Just ask the yardman for broken stones. They should sell for about

CUTTING FLAGSTONE

1 Use the corner of a stonecutter's chisel or any other tool to score a straight line across the stone.

2 Lay the stone on top of a board, with the score mark centered over the middle, and rap the chisel lightly and continuously, sliding it along the line. When you have cut a groove, turn the stone over and cut a second groove opposite the first.

3 Move the stone so the line of cut is about 1 inch beyond the edge of the board. Then tap the end of the stone lightly with a small sledgehammer until the stone breaks along the line of cut.

half price. Stay away from yards that call broken stones "random" and charge as much or more for them than for unbroken stones. This is like paying a premium for wood with knots just because it is called knotty pine. Bear in mind, when purchasing broken stones, that if you are going to lay them on soft soil, pieces smaller than 1 square foot or so are of no use. Also, you may have to do considerable cutting to make the broken pieces fit, if you want to hold the edges of your walk to a straight line.

FIELDSTONE WALKS

An even less expensive, though not easier, way to make a walk is to use field-stone, assuming there are enough suitable stones to be found on your land or that of your neighbor. For use in a walk, it is essential that the stones have one or more flat sides. Many fieldstones are glacial debris, stones that were pushed along by glaciers as they inched southward. These have been well rounded by their journey. If you have a sledgehammer, try giving the round stones a sharp rap on their sides, especially if you can see any kind of line there. Often, the stones will split flat across.

As before, cut the grass as close to the earth as you can along the path your walk is going to take. Then sweep the path clean. Take one of the stones you have gathered and place it flat side down on the ground. Scratch its outline in the ground with the point of a trowel. Remove the stone and dig a hole within the outline that conforms as closely as you can readily manage to the rounded side of the stone, but about 2 inches shallower. Drop the stone, round side down, into the hole. If you have judged your excavation correctly, the top of the stone will be level and project about 2 inches above the ground. If the stone is too low, remove it, toss some dirt into the hole and try again. If the stone is too high, inspect the hole for compression marks—where the stone pressed into the dirt. Dig there a little and try again.

When you have positioned all the fieldstones along the desired course of your path and about an inch or two apart, pick up the pieces of sod (grass and roots) you removed making the holes. Cut them as necessary and fit them, grass side up, between and alongside the stones. Tamp the sod into place. Sweep the stones clean and you are done.

Walks have been paved this way for countless centuries. They never wear out, but the stones do sink into the earth. When this happens, just lift them up, spread some soil or sand under them as necessary and lay them back in place. That will hold them for another five to ten years.

LAYING FIELDSTONES

1 Lay one stone on the ground and cut its outline into the sod with a trowel or shovel.

2 Remove the stone and dig a depression within the outline that conforms, as closely as possible, to the bottom of the stone. Replace the stone.

3 Repeat the same steps to position each stone in the walk.

WORKING WITH SMALL FLAGS

If you want to pave your walks with small flags, say concrete block or brick, you have either to accept the unevenness that will result should you lay them directly on cut grass (because grass is never perfectly flat), or excavate and otherwise prepare the surface of the path. There are two ways to do this. We will tackle the easy way first.

Laying on grass. Start by cutting the grass along the desired path and cleaning up the debris. Then spread a layer of sifted soil or sand over the surface of the path. Wet the sand or soil lightly, then level the surface by dragging a straightedged board across it.

When you have made the surface of the path reasonably level and smooth, you can lay the small flags directly on top. Unfortunately, the flags will not remain level for very long. How long is difficult to estimate. However, it should be safe to assume that it will be at least a year before the grass decays and the earth compacts sufficiently to necessitate lifting and repositioning the flags.

To lay small flags on uneven ground without excavating, cut the grass short and spread a layer of sifted sand on top. Level the sand with a straight board and use string guides to align the flags.

LAYING FLAGS ON EXCAVATED SOIL

1 Remove the sod with a long-handled shovel.

2 Level the bottom of the excavation or trench. A square-edged shovel is best for this job. The board across the excavation serves as a depth guide.

Using an earth base. This method requires more effort and additional expense, but the result will be a smoother, flatter surface that should hold that way for many years.

Start by marking the course of the proposed walk with mason's lines (strong strings) stretched between pegs driven into the earth. Position the lines parallel to one another and about 1 foot wider apart than the desired width of the walk. With a shovel, carefully remove the sod from between the lines to a depth of about 2 inches. Make the bottom of the excavation as smooth and level as you can. Then with an iron rake loosen the soil a bit and free it of stones. Since it is customary to make the top surface of a walk about 2 inches above grade (i.e., above the surface of the soil), you will have to use 4-inch-thick flags—for example, bricks laid on edge or standard,

14

3 Rake the bottom of the excavation free of stones and loosen the soil.

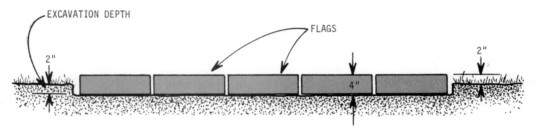

EXCAVATION DEPTH

FLAGS

2"

2"

4"

4 Lay the flags directly on the excavated soil, as shown in this cross-sectional view.

4-inch concrete block. These and similarly thick flags can be laid directly on the soil in the shallow trench in a variety of patterns.

Check the first flag you lay down with a spirit level. If it isn't level, or following the natural pitch of the path, add or remove a little soil from beneath it as necessary. Then place a stick between the first and second flags to space them easily and accurately if they are not to touch each other. Use a level on top of a length of 2x4 to make certain the second flag is flush with the first. If it is high, pound on top of the wood a little; if low, add some soil underneath. If you take the trouble to check every flag as you lay it down you will end up with a fairly flat path, and possibly a stiff back if you try to do the whole job in one afternoon. There is no reason to work too hard; rest before you feel strained.

You can, if you wish, add filler (soil or sand) between the flags as you go, or you can wait until they are all in place and then spread the filler over all the flags and brush it into the spaces. After filling the joints, bank the filler alongside the flagging. This keeps the flags in place and prevents people from stubbing their toes on the edges of the flagging.

LAYING FLAGS ON SAND

If you want your walk to be especially smooth and level, or if you want to use flags less than 2 inches thick, lay the flags on a sand base. Starts as before and lay out the path with parallel lines spaced 1 foot wider apart than the width of the walk. Excavate as before, only this time go down about 3 inches, taking care to keep the flat, wide trench bottom fairly level and smooth. Digging this deep gets you past much of the topsoil and roots and into the subsoil, which is firmer and not as likely to compact.

Next, install a pair of guides lengthwise in the trench. Use 2x4s or 1x4s. Place them on edge, parallel to one another and just as far apart as the proposed width of your walk. Drive stakes into the ground next to the outsides of the guides. Nail the guides to the stakes, making the top edges of the guides just as high above grade as you want the finished surface of the walk to be. If one pair of boards isn't long enough, install additional pairs. Then cut the stakes flush with the tops of the guides. To curve the walk, outline the curve with stakes. Then bend and nail ¼-inch plywood strips to the stakes. The curved guides serve the same purpose as the straight ones.

The next step is to fill the trench with some type of granulated base material which can be sand, cinders, crushed stone, gravel, or a combination of these. The purpose of the granulated base is to provide drainage and a perfectly smooth surface on which the flags will rest. The best arrangement is a layer of crushed stone topped by a layer of sand an inch or so thick. Do not use crushed stone or gravel alone, if you have any choice, because they are difficult to work with.

The height to which you fill the trench depends on the thickness of the flags you have selected. If, for example, you are going to lay down 4-inch-thick block and the excavation is exactly 3 inches deep, 1 inch of sand will bring the block's surface 2 inches above grade. If you are laying 1-inch stone, you would, of course, need 4 inches of sand to bring it 2 inches above grade.

Incidentally, neither the 3-inch trench depth nor the 2 inches of height above grade are sacrosanct. You can vary these dimensions any way you wish. However, if

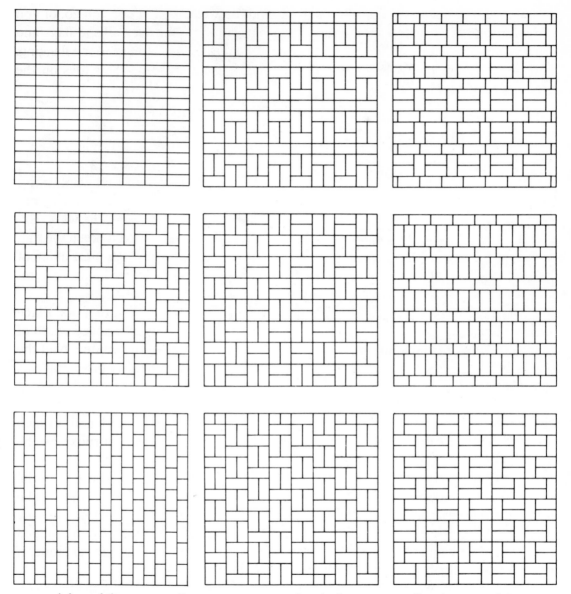

A few of the many patterns you can use when laying common flagging material.

there is much more than an inch of sand base above the ground, small flags tend to shift underfoot until the sand along the edges of the walk has been stabilized by grass.

Once you have decided on how many inches of granular base you need, the next step is to estimate the quantity to purchase. To do this, simply multiply the total

17

A simple but interesting walk made of Belgian blocks laid directly in the soil with sand between the stones.

width of the trench — the base material should spread out past the guides — in inches by the desired height of the base, again in inches, by the length of the walk in inches and divide by 1,728 (the number of cubic inches in a cubic foot) to get the number of cubic feet necessary. Then divide by 27 to convert cubic feet to cubic yards, which is the measure generally used when ordering sand, crushed stone, and other such materials. (All this math is easy if you use a pocket calculator.) Then add 20 percent to allow for variations in the actual depth of the trench and waste.

Next, place a stop board across each end of the walk. The board stops the sand you are now going to dump into the trench from sliding out past the guides. Spread the sand with a shovel and then rake it level.

LAYING FLAGS ON SAND

1 Dig a trench 3 inches deep and 1 foot wider than the desired width of the walk. Install a pair of wooden guides within the trench, using stakes to hold the guides in place the exact width of the walk. As always, the depth of the trench and the height of the guides depend on the thickness of the flags. In this walk bricks are to be used, on edge.

2 Nail the guides to the stakes using a hammer or rock as a backstop.

3 Drive the stakes into the ground to bring the top of the guides 2 inches above grade.

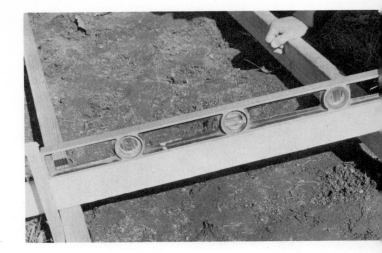

4 Level the guides or give them a pitch if drainage is a problem.

5 Dump gravel into the trench . . .

6 . . . and spread it with a rake.

7 Lay a straightedge across the guides and check the depth of the gravel so it is uniform throughout.

8 Cut the stakes flush with the top of the guides. Then dump sand in the trench and spread with a shovel or rake.

9 Make a strike board (see drawing next page), and smooth the sand to the correct height.

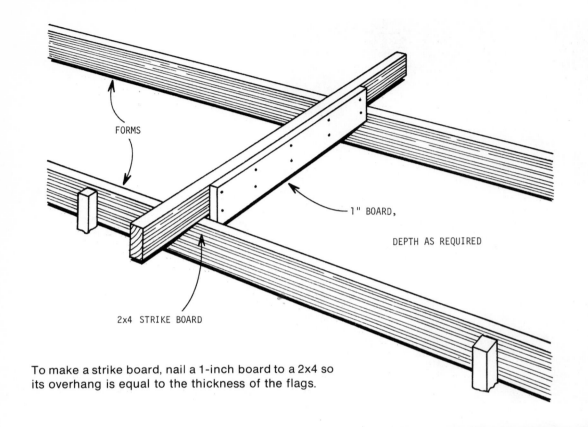

FORMS

1" BOARD,

DEPTH AS REQUIRED

2x4 STRIKE BOARD

To make a strike board, nail a 1-inch board to a 2x4 so its overhang is equal to the thickness of the flags.

10 With the sand level, smooth and at the desired height, lay the flags, or bricks in this instance, in place. Space the bricks by eye or use a board spacer to help you.

11 To flatten the surface of the walk, pound the bricks into the sand with a block of wood and a small sledgehammer.

12 Finish the walk by sweeping sand between the bricks.

Now construct a strike board (not to be confused with a screeding board), as shown in the accompanying drawing. With a helper holding the farther end of the strike board, push the board down the length of the walk. As you push the strike board along the guides, its bottom edge strikes off the surface of the sand and produces a level surface at exactly the desired height.

A flagstone walk can be finished by troweling sand, or a damp sand/cement mixture, into the joints. Use the point of the trowel to pack the sand/cement mixture (mortar) tightly and smoothly into the joint.

This done, you can either wait for a light rain or speed up the process by wetting the sand with a fine spray. Damp sand is easier to work with.

Next, lay the flags down as previously described. Use a 2x4 long enough to span two or more flags and with a small sledgehammer tamp them into the sand, bringing each flag flush with its neighbor. Place a spirit level on top of the 2x4 now and again to check the levelness of the flags that are in place. In addition, keep your eye on your overall progress by placing another, longer 2x4 across the two guides. Use this board as a height gauge and keep moving it along to make certain the top of your walk doesn't wander uphill or down as you work. It is difficult to hold a long surface perfectly true with a level alone.

Finally, pull the guides and bank soil against the edges of the walk. If you have left spaces between the flags, they have to be filled as already described.

When the guides have been removed, fill the space between the walk and the grass with pieces of sod you removed earlier.

FLAGGED PATIOS

Besides adding a touch of elegance to any home, a patio provides an area for relaxation and entertaining that is easily cleaned and impervious to the weather. Like a dry flagged walk, a dry flagged patio is the easiest and least expensive of all types to construct. In addition, dry flagged patios are flexible. They are not damaged by frost or other movements of the earth. A 20-by-30 foot patio can be completely paved this way in a single afternoon for little more than the cost of the flags themselves. The appearance of this type of patio in no way betrays its ease of construction and low cost. In some ways the dry flagged patio can be more attractive than the conventional — and more expensive — types of patios.

The construction of a dry flagged patio is similar to that of a flagged walk. Think of a flagged patio as consisting of a number of flagged walks side by side and you have mentally pictured most of the job.

LAYOUT

Free-form patios are constructed of square or rectangular paving laid down more or less as it comes to hand without thought of spacing or edges. The flags just blend into the grass.

This formal patio was made of individually cast concrete slabs, an ideal do-it-yourself project. A strip of sod 6 inches wide separates the slabs.

Patios paved at random are paved with odd-shaped (broken) flags which are spread over the area, an effort being made to hold the spacing between the stones more or less to one dimension. In a sense, a randomly paved patio is put together somewhat like a jigsaw puzzle.

Formal or regular patterns are most easily and least expensively produced by using flagging in available sizes and grouping a few stones in a simple pattern that is repeated a number of times to produce the overall patio pattern. When you work this out on paper, remember the stones are usually a little undersized and allow for spacing.

Although it is a matter of taste as to which design is preferable, free-form, random, or formal, one characteristic of these designs is definite; they cannot be mixed. The result will be visually poor.

To lay out a random design with a minimum of labor and waste, spread the flags temporarily over your lawn and shift them about until you are satisfied with the results. Number each stone with water-soluble paint. Copy the pattern on paper and number each stone in the drawing. Then you can move the stones to their final position in the patio area without danger of mixing them up.

If the sizes of the available flags do not suit your purpose, you can cut them yourself or have the yard cut them to your sketch. Their fee includes breakage, which is inevitable, and the small useless pieces that usually result from some of the cutting.

Flagstones should be stored on edge to protect them from breaking.

Patio made of Belgian blocks laid on end, in concentric semicircles, directly in the soil.

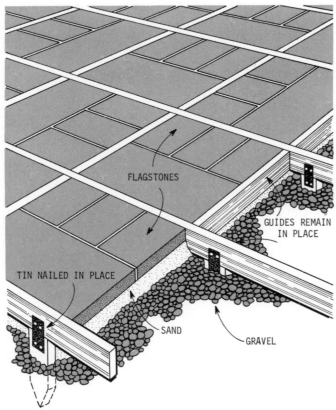

FLAGSTONES

GUIDES REMAIN IN PLACE

TIN NAILED IN PLACE

SAND

GRAVEL

Wood dividers can be used to hold flags in place. The dividers can be arranged to form interesting patterns.

WORKING WITH LARGE FLAGS

Large flags are best for paving patios because they require less handling and can be laid directly on closely cropped grass. Since flagging is usually sold by the square foot, large flags cost no more than small ones and sometimes a little less.

First, cut the grass as closely as you can over an area a foot or more larger than the perimeter of the patio. Clean the space of all grass clippings and small stones.

When A = B all sides are parallel and all corners square.

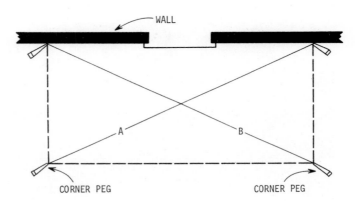

When A, B, and C are in the ratio of 3, 4, 5, the corner is square.

WHEN THE RATIOS ARE:

A	B	C
3	4	5
6	8	10
9	12	15
12	16	20

THE ANGLE IS 90 DEGREES

Use pegs and lines to lay out the rectangle which will form the perimeter of the patio. Check the measurement of each side with a steel tape. Then measure the distance between each pair of diagonal corners. If the distance between these pairs of corners is the same, all four corners are perfectly square. If one side of the patio is parallel with a wall or building, then the patio itself is square with that structure.

Next, lay one row of flags alongside a border line, using a stick of the proper thickness as a spacing aid. Now stretch a second line parallel to the border line and use that line as a guide to laying the following row of flags. When all the flags have

Constructing a patio by laying large flagstones directly on grass. When the pattern is completed, sand is brushed between the flags.

A smooth-surfaced patio can be made by laying flags on a base of sand that has been spread over the grass.

been properly positioned, fill the spaces between them with sand or loose soil and grass seed. Bank sand or soil along the edges of the flagging and the job is done.

Should you wish the patio surface to be smoother than it is possible to make it by simply laying flags on cut grass, here is an easy way to achieve this. Start as before. Cut the grass and clean the area. Then spread a layer of sand or soil over the grass. You need little more than enough to cover the surface. Next, with the help of a friend, drag a long, straight board across the ground a few times until you have leveled the patio area. Now you can lay the flags down exactly as before, but the result is a smoother patio surface because the sand or soil has filled in the irregularities in the earth and provides a stabler, smoother base.

FIELDSTONE PATIOS

Fieldstone patios are made almost the same way as fieldstone walks. You simply cover a larger area.

Start as before by cutting the grass closely over a slightly larger area than that which you plan to flag. Mark the perimeter of the patio with a mason's line. Dig a little hole for each stone and insert the stones into the holes, flat side up. The depth of the holes should be such that the surface of the stones is about 2 inches above the soil. Space the stones about 2 inches apart; more if you plan to use grass between them.

Use the turf you have removed in the process of digging the holes to fill the spaces between the stones and to form a little ramp of sod along the edges of the pavement.

Making a patio of fieldstones. Stones are placed flat side up directly on the grass, or in individual excavations fitted to the underside of each stone. A spirit level on top of a long, straight board helps to keep the surface of the patio level.

Grass sod is used to fill the spaces between stones and to form a gentle slope leading to the patio.

A fieldstone patio finished by sweeping sand between the stones.

Mortar joints. There are several ways to join the edge of one stone or concrete flag to another. The method described below is perhaps the best to use with field-stones. Other methods are described in Chapter 15. Bear in mind that the mortar you place between any kind of flagging laid directly on the ground will crack with time and frost. The cracks can, of course, be patched and in fact do no harm to the patio.

Using a small shovel or trowel, mix 1 part regular cement with 4 parts sand and pour the dry mixture into the spaces between the stones. Where there are small depressions along the edges of the stones, fill them too. Then adjust your garden hose to an extra-fine spray and slowly and carefully wet all the joints, taking care to soak them, but not to wash the mortar away. Wait a few days for the mortar to harden before walking on the patio.

WORKING WITH SMALL FLAGS

It is best not to lay flags less than 1 foot wide and less than 1½ inches thick di-rectly on grass or even on grass covered by a thin layer of sand or dirt. Small flags are difficult to lay smoothly on grass. They also turn underfoot. In addition, when the grass rots, as it inevitably must, you have the job of lifting the small flags, placing sand under them and replacing them. This isn't much of a chore with large flags, but with small flags it can be twice as much work as constructing the patio in the first place. Moreover thin flags will crack if not properly supported.

To prepare a suitable base for small and/or thin flags, use pegs and line to mark the perimeter of an area 1 foot wider and longer than the patio. With a shovel, remove the grass and its roots. Work carefully. Keep the bottom of the shallow ex-cavation as level and as smooth as you can.

The next step depends on the thickness of the flags you plan to use. Assuming the flags are 4 inches thick, which is the nominal thickness of standard, solid concrete block, or a little more than the width of common brick, your excavation needs to be 2 inches deep (which is probably about where you are after sod re-moval). If you lay 4-inch-thick flags down now they will project about 2 inches above grade. On the other hand, if you plan to lay Belgian blocks down, you have to dig down another 2 inches to make the finished surface of the patio 2 inches above the surrounding ground. As stated before, you can make the surface higher or lower as you wish. The 2-inch height is merely the usual one.

Assuming that you are down 2 inches and are going to lay 4-inch flags, prepare

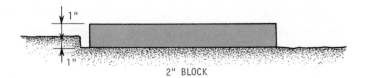

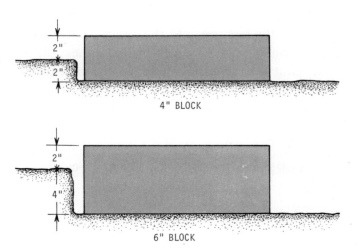

2" BLOCK

4" BLOCK

6" BLOCK

Relationship between thickness of flag and depth of excavation.

Flagging a patio with 2-inch concrete blocks on soil. After removing the grass, rake the soil level and lay the stones (right). Remove or add soil, as necessary, with a trowel. Use a small sledge and a length of 2x4 to bring the surfaces flush and check them with a level (opp. page).

the bottom of the excavation by raking it to loosen the soil and remove stones. Next lay the flags in place as previously described. They can be spaced apart. They can touch one another. They can be positioned at random. They can be laid according to a simple or intricate pattern. The basic procedure remains the same. The flags are kept level and flush by using a 2x4 and spirit level. Add or remove soil from beneath the flags as necessary, and pound down those that are only a little high.

USING A GRANULAR BASE

When you want to use thin flags or you want to secure a smoother surface than can be achieved by laying flags of any size and thickness directly on the soil, lay the flags on a granular base. This can be sand or fine cinders, alone or laid over gravel or crushed stone.

Start as before, but excavate to a depth of 3 inches, again taking care to keep the

bottom of the excavation as flat and level as you can. Going down the extra inch gets rid of a little more topsoil, and so reduces the degree of compaction and consequent sinking of the flags — another advantage of using a granular base.

Next, install a pair of 1x4s or 2x4s nailed to stakes to serve as guides along two sides of the excavation. Space them just as far apart as the width of the patio. Their top edges should be exactly as high as you want the surface of the flags to be. Level the guides. Flagged, flexible patios normally do not require a pitch, since rainwater seeps down between the flags.

If the two guides are more than 10 or 12 feet apart, install a third guide between them, its top edge level with those of the two other guides. To check this, place a long, straightedged board across all three guides. If you don't have a long enough board, stretch a line across the three guides and adjust the height of the center guide as necessary. This done, cut all the stakes flush with the tops of the guides.

As when making a flagged walk using a granular base, your next step is to calculate how much granular material to order. You need just enough base to bring the surface of the flags flush with the tops of the guides. Generally this elevation is 2 inches above grade. So you need 1 inch of sand or similar material with 4-inch flagging, 2 inches of base material with 2-inch flagging and 4 inches of base material with 1-inch flagging.

When you need more than 1 inch or so of granular base it is advisable to use gravel or crushed stone on the bottom and top it with 1 inch of sand. The coarse material drains better than the sand and also provides a more stable support. However, do not use gravel or crushed stone alone; it is much too difficult to work with.

The easy way to estimate the quantity of base material needed is to figure everything in inches. Multiply the length of the excavation by its width by the desired base thickness, all in inches, using a pocket calculator. Then divide the result by 1,728 (the number of cubic inches in a cubic foot) to get cubic feet. Divide this figure by 27 to get cubic yards, because you order sand and crushed stone by the cubic yard. Finally, add 20 percent to the last figure to take care of losses.

If the base material is to extend above grade you will need stop boards to keep the sand from spilling too far beyond the ends of the guides. Make the stops from 1x4 or 2x4 lumber. Place the boards on edge and drive stakes behind them to hold them in place. Nail the stop boards to the stakes with the tops of the stops flush with those of the guides.

Spread the granular material over the excavation. If coarse material is used, it goes down first, followed by the sand or cinders. Level both layers with a shovel and rake.

FLAGGING A PATIO ON A SAND BASE

1 The perimeter of the patio is demarked with pegs and string, leveled with a line level. The height of the string is measured at several points to determine whether or not the ground is level, and if not, how much correction is needed. The soil is then excavated to the proper depth, in preparation for laying a sand base.

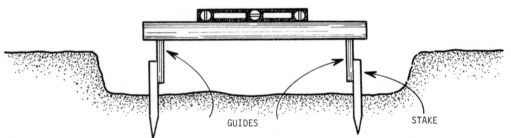

GUIDES   STAKE

Cross-sectional view of a patio excavation showing the position of the guides and the use of a straightedge and spirit level to level them.

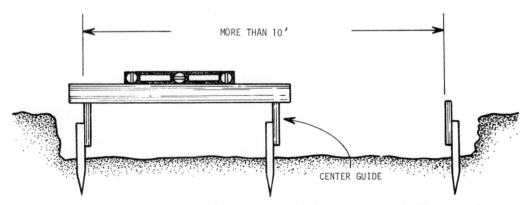

MORE THAN 10′

CENTER GUIDE

When two guides are more than 10 feet apart, a third should be erected between them.

2 After filling the excavation with sand, level and smooth the surface with a strike board.

3 Lay the first flags against one edge guide and level them. If necessary, hammer a high flag lower, or remove sand from beneath the flag.

4 Use a long, straight board to keep flags aligned with center guide.

5 After removing the center guide, use a trowel and a short board to fill the opening and smooth the sand.

6 Lay the rest of the flags in place and align them with a string.

7 To finish the flags-on-sand patio, sweep sand into the joints.

41

Fashion a strike board from a 2x4 long enough to straddle two guides and a 1x6, 2 or 3 inches shorter than the distance between two guides. Nail the 1x6 to the 2x4 so the 1x6 projects over the edge a distance equal to the thickness of the flags. If you are going to use 1-inch flags, for example, nail the 1x6 to the 2x4 so it projects exacly 1 inch beyond the edge of the 2x4.

Now the strike board is placed across two guides. The 1x6 projects downward, in this example, exactly 1 inch. Push the strike board, preferably with an assistant on the other end, down the length of the patio. As you move the board its bottom edge will level the sand to within 1 inch of the top edges of the guides. (No doubt you will have to do a little raking and shoveling as you go along to help the strike board.)

If you have three or more guides you will have to repeat the striking operation. When done you will find a small, long mound of sand beneath the middle guide. Let this be for the moment.

Finally, soak the sand or cinders with a fine spray from a garden hose. This will make the granular material easier to work with.

At this point you have a slightly damp, perfectly smooth and level layer of sand lying on top of the earth exactly where you want your patio to be. The sand is surrounded on two or more sides by wooden guides. The distance between the top edges of the guides and the surface of the sand is exactly equal to the thickness of the flags you are going to lay down. All you need do now is place your flags within the guides.

Starting at one of the corners, lay the first flag down. It may touch the wood guide lightly. Check the top of the flag against the top edge of the nearby guide; which establishes elevation. If the stone is a bit high, lay a piece of 2x4 on it and try pounding it down a little. If that doesn't do the trick, lift the stone, remove some sand and try again. If the stone is too low, add a little sand under it. As you continue to lay flags, check the surface of one against the other with the 2x4 and a spirit level.

As you work progressively farther from the guide you will need a longer 2x4 to make certain the flags are level with the guide and the rest of the stones. In addition you may find it helpful to stretch a line parallel to a guide to check on the alignment of the stones.

Continue flagging this way, checking the rows for alignment and height as you go until you are close to a middle guide, if there is one. Now is the time to pull this guide out. Use a trowel or small shovel to remove the excess sand and fill the holes left by the stakes. Then slide a short, straight stick across the sand to level the area where the middle guide was placed. Now you can continue flagging until you reach the last guide.

When the last flag is in, check them all once again for height and alignment. Fill the spaces between the flags. Pull the guides and stops. Bank soil or sand up against the sides of the patio.

ADVERSE CONDITIONS

Since it is unlikely everyone will have a level, dry yard that is easily paved, we have a "problem" section.

Wet ground. Assuming the area is level and that water lingers on after a rain because of a high water table or the amount of clay in the soil, the simplest solution is to lift the patio a foot or so above grade. Do this by omitting the excavation and installing guides a foot or so high. The area between the guides is first filled with coarse base material and topped with an inch or so of sand. It is then flagged as described. When the guides are removed, sufficient gravel topped by sand or soil is placed alongside the patio's edges to make a gentle slope. In this way the height of the patio can be more or less hidden.

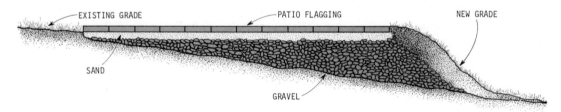

When the natural slope of the ground on which you plan to build is more than ½ inch to the foot, correct it by building a terrace of soil, crushed stone, or gravel.

Excessive slope. Any of several steps can be taken to correct excessive slope. One is to cut into the high side of the hill and if necessary to place a small retaining wall there. Another is to build up the low side by placing fill (soil) or gravel there. A third method combines the first two. Some soil is removed from the high side and laid over crushed stone or gravel on the low side to make one level surface.

ORNAMENTAL STONE WALLS

A stone wall is an informal border that can mark the edge of your property. Beyond that its existence is almost entirely ornamental. True, it can provide homes for chipmunks and mice and safe places for squirrels to hide nuts and it can support rambling roses and blackberry bushes, but a stone wall will not keep dogs and children out. When you build a wall by piling one stone on top of another you are building a memento of a slower and supposedly happier time gone by, and for the very pleasure of building.

Perhaps surprisingly, you will find building a stone wall very satisfying if you do not rush and if you take care not to lift stones beyond your capacity.

DRY STONE WALLS

Dry stone walls are constructed like flagged walks and patios, completely without mortar, which makes them flexible, easy to construct, inexpensive, and durable. Since the wall is flexible it can rise and fall with frost heave without cracking. These walls need no foundation of any kind. Actually, you can build the wall right on top of the grass.

You can use any size, kind, shape, and color of stone available. You can even use pieces of concrete if you wish. The very best stones are naturally those you get free. The next most desirable quantity is flatness. The flatter the stones, the straighter you can hold the sides of the wall. The worst stones are "baldies," which are smooth, more or less round boulders. When you are stuck with baldies you have to pile them pyramid fashion, just like cannon balls, which makes for a triangular cross section.

The bigger the stones, the easier they are to work with, up to a point, and that point is determined by how much you can *easily* lift. Above all, avoid straining yourself. If your maximum safe lift is around 100 pounds, which is about the maximum for a grown man in middling shape, you will soon tire lifting 100-pound stones. With stones weighing no more than 30 pounds you will be able to work all day.

In addition to as many large stones as you can secure, you will need an assortment of smaller stones, unless the large stones are of equal size and fairly regular. This, of course, is rare, unless you purchase quarry stones cut to size. The small stones and stone chips are needed to fill some of the spaces between large, irregular stones.

You may, like the farmers of New England, find yourself blessed with an abundance of fieldstone, which is wonderful only if you are not going to farm. But if you are not this lucky, you can sometimes find stones for the asking by visiting building excavations and roads under construction. Generally the contractors will be happy to let you save them the trouble of hauling the stones away.

When construction sites are unavailable, look to urban renewal and demolition. Lots of old buildings rested on rubble masonry foundations. Since builders were generous with sand in those halcyon days, the mortar used to hold foundation stones together crumbles very easily. So if you can get to an old foundation, you should have no trouble removing all the stones you need.

As a last resort, purchase rubblestone from the nearest quarry. Rubblestone consists of the odd-shaped pieces of stone left over after the quarrymen have blasted and picked out the better shaped stones for cutting and trimming. Rubblestone is the cheapest you can buy. Get the truck-load price and the carry-it-home-yourself price.

If, however, you want a more formal stone wall, in which the stones are more or less the same size and have that "fitted" look, you are going to have to purchase cut stone. Each quarry has its own designations, with the stones divided into specific categories with specific prices. The more carefully the stones are selected and cut, the more uniform they are in color, size, and shape and the higher the price. They are also the easiest to use. Perhaps the best compromise between appearance, cost, and

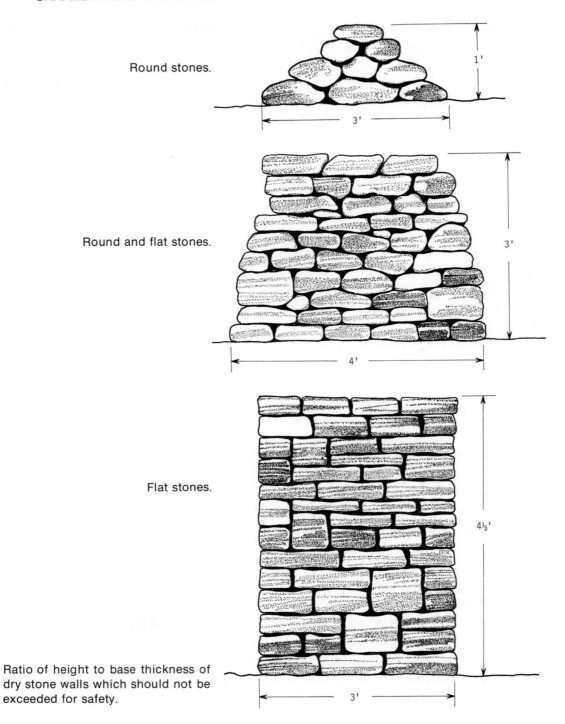

Round stones.

Round and flat stones.

Flat stones.

Ratio of height to base thickness of dry stone walls which should not be exceeded for safety.

labor is the category usually called veneer. These stones are relatively flat, about 4 to 6 inches thick and from 1 to 3 feet in length. They may cost several dollars per square foot.

Since you are not going to use any cement to glue the stones together, their weight alone must hold them in place. Therefore, you cannot safely make any dry stone wall much higher than it is wide even when using the best of all possible stones.

Basic principle of dry stone wall construction is to overlap stones, both lengthwise and crosswise. This locks them in place and produces a strong wall.

Going by the rule of the ancient mason's thumb, you can make a dry wall one and one half times as high as its base width when using really flat stones. For example, a dry stone wall 3 feet thick at its base should never be more than 4½ feet high.

When using moderately flat stones, hold the wall's height equal to its width. Thus to make a 3-foot-high wall with such stones you have to start with a 3-foot-wide base. We specify the base width because in many instances when working with less than perfectly flat stones you will find yourself making the wall a bit narrower with each course (layer) of stones you put into place. It is not unusual to start out with the wall 3 feet thick at its base and end up with a thickness of no more than 2 feet at the top.

When using either stones that aren't flat or a mixture of flat and round stones, you have to limit the height of your wall to about three fourths of its width to keep it safe and stable. Therefore, if you want to erect a 3-foot-high wall you have to start with a base that is 4 feet wide. When you are working with unevenly shaped stones, the width of the wall along its top will probably be no more than 2 feet.

A dry stone wall made of fairly flat stones set close to one another is approximately 20 percent air. So if you figure the cubic volume of the wall (height times width times length, all in feet) and order this quantity of *cut* stone, you will be fairly close.

On the other hand, if you are going to use quarry rubble, be forewarned that the rubble is merely dumped into the back of the truck, which results in a load that is one third or more air. When you build a wall you are forced to position the stones much closer to each other. Therefore when ordering quarry rubble it is advisable to order at least 20 percent more than the volume of the wall you plan to build.

Building the dry-stone wall. Start by marking one side of the wall by either laying down a long board or by stretching a line between pegs. Then spread out the stones that you are going to use in a single layer 3 feet or more away from the planned front side of the wall. Spreading the stones out this way makes it easier to select the ones you need. The 3-foot space between the stones and the front of the projected wall gives you working room. Now position a second guide parallel to the first to mark the back edge of the wall.

Some of the old-time New England wall builders were reputedly so fast they could keep an unwanted stranger from walking on their property by just building a wall in front of him, but we have slowed down since then. Laying a 10-foot stretch of a 3-foot-high wall in an afternoon is a fair rate of achievement for anyone who has never done it before.

Use stones that are too heavy to be easily lifted for the bottom course of the wall. Position the stones that you select for the first course in a single row just within the line or board that marks the front side of the wall. Adjust each stone so that its flattest side is on top and level, and its next flattest side faces outwards. Use small stones and/or stone chips to hold the larger stones upright and immobile. Repeat this operation along the rear side of the stone wall, taking care not to place the large stones forming the second row and rear edge of the wall directly behind those forming the front row. Next fill the space between the two rows of stones with small stones and chips to a height that is level (or as level as you can make it) with the tops of the large stones.

The next step consists of laying down a second course of stones. Since our model wall is two rows thick, this course again consists of two rows of stones, which must be positioned above the spaces between the stones in the lower course. In other words, the first and second layers of stones must be staggered. Again you must take care either to select odd-shaped stones to fit odd-shaped spaces or to use small

1 Mark the front edge of your wall-to-be with a line or, as here, with a long board held by stakes.

2 Place a stone against the guide board, next to the stake that marks the end of the wall, with the best side of the stone facing upward, the next best outward. If necessary, use a small stone to level large one.

49

3 Mark the rear edge of the wall with a second guide board and also an end stake; then lay a row of stones along the inside of the second guide.

4 Fill the spaces between the front and rear rows of large stones with small stones and stone chips.

5 Lay a second course of stones on top of the first. If possible, use stones large enough to span the two lower courses and level them with small stones. (*Continued on next page.*)

Using a chisel and a small sledgehammer to split a stone.

6 Top the wall with the largest and flattest stones. Remove sharp, projecting points with a small sledgehammer. Do not drive small stones into the wall; this will loosen the structure.

stones and chips to hold the tops of the second course of stones level. Next, the space between the two rows of stones forming the second course is filled with more stones and made level.

At this point you have two courses of stones, with the stones staggered in the upper and lower layers in both the front and rear rows. Now you have to lock the front rows to the rear rows by laying down a third course of stones which are large enough to reach from the front to the rear of the wall. Since you most likely will not have stones this large and flat, unless you have opted for quarry-cut stones, do the best that you can. Lay the third course down the middle so that each stone overlaps as much of the front and rear stones as possible. Remember, it is this constant overlapping of courses that holds the dry stone wall together. Next you can lay two rows of smaller stones on either side of the middle row.

From here on up this three-course pattern is simply repeated as many times as necessary.

The wall is usually topped with the largest and smoothest stones available. However, you can also terminate it with whatever stones naturally end up on top. The wall will be just as strong if you do so. Another possibility, if you would like the top of the wall to be relatively flat but do not have the necessary stones, is to use mortar. Mix 1 part cement with 4 parts sand. Add sufficient water to produce a consistency similar to ice cream. Then using a trowel, or even a small piece of wood, spread the mortar in a smooth, 2-inch layer over the top of the wall. This will help hold the stones in place and prevent children and dogs from spoiling your handiwork.

When you have to cut stone, look for the grain, and if there is any, rap the stone along the grain with a small sledgehammer. If that doesn't do it, try the hammer and a chisel on the grain.

STONE AND MORTAR WALLS

Walls and similar structures constructed from mortar and stone differ from dry-stone work in three principal ways: Firstly, mortar-and-stone walls are inflexible. If you do not want them to crack in winter they have to be provided with a base that reaches below the frost line. Secondly, they may safely have perfectly vertical sides no matter what type or size of stones is used. And lastly, these walls are so strong they cannot be taken apart with anything less than a sledgehammer.

Many stone-and-mortar walls are made of cut stones that are neatly trimmed and fitted, but that doesn't mean that such walls are the most attractive. Style in walls is a matter of taste and surroundings. To some, a country home is best set off by walls of irregular fieldstones. Some prefer wall surfaces that are almost flat. Others choose stones with character, and it is not uncommon for masons to give stones such character by trimming their edges back a half inch or so. This results in stone surfaces that project a half inch or more beyond the surface of the mortar joints.

To decide what kind of stone to use, look at various walls until you find the design you like. Then study it carefully because many things go into its appearance; the size, shape, type, and color of the stone, the width of the joints, the mortar treatment, and so on. The second step consists of visiting the nearest quarry to learn what stone they have and what they charge for it. Rubble is the least expensive quarry stone. The next is usually veneer, which is stone selected for one dimension only. In most cases veneer is between 4 and 5 inches thick, but the other dimensions are unselected. The next, more expensive, grade is probably selected veneer, followed by a grade which includes some cutting. Naturally, the more trouble taken in selection and cutting, the higher the cost. So it is advisable to check on stone costs before you commit yourself to a particular type of wall.

In any event, no matter what type or shape of stone you select, the procedure for working with mortar and stone remains the same.

It is difficult to estimate accurately the quantity of stone you will need for a particular job. The best suggestion this writer can make is that you compute the cubic volume of the project and increase it by 20 percent to determine the quantity of stone you should order.

An attractive roadside wall made of fairly large boulders with recessed mortar joints.

A partially mortared stone wall built mostly of large "baldies." The mortar that joins them can be seen in only a few places.

This wall was made of only rectangular stones held with mortar. The wide joints were roughly raked.

Selecting a stone from a pile of veneer dumped by the delivery struck. Note the sheet of plastic to keep the stones dry.

Preparing the mortar and stone. To make mortar for use with stone mix 1 part *regular* (not mortar) cement with 3 to 4 parts sand. Mix the two dry and then add sufficient water to make a mix with the consistency of ice cream. Generally it is best to have the mortar a little on the dry side. No lime is added; that is why you do not use mortar cement, which contains lime.

To ready the stones for joining, wash and scrub them dish clean. If there is any dirt on them at all or any moss or dust, the mortar will not bond to the stone and there will be no joint. When you have finished scrubbing the stones, let them dry and cover them to protect them from rain. If you apply mortar to a wet stone the mortar will soften and the stone will slide out of place, which will make the job that much more difficult.

Cutting stone. Cutting and trimming stone is an art, which means that the following instructions will only start you on your way. It will take time before you get the feel of the stone and are able to cut it easily and accurately.

The stone to be cut is placed either on the ground or, if you do not wish to bend, on a 4-inch-thick bed of sand held in an open frame on a sturdy table. If you place the stone on a concrete slab or another stone, it will bounce when you hit it and will most likely shatter.

Use a small sledgehammer and a mason's chisel. When the edge of the chisel wears off, grind it to a 75-degree bevel. The chisel is used to chip a line which weakens the stone at that point. Generally a stone can be cut most easily along its grain, but the grain in dense stones is difficult to decipher; it requires experience.

To reduce the length of a stone by, let us say, an inch, place the stone flat on its side. Position the chisel vertically on the desired line of cut and rap the chisel hard, again and again, taking care to hold its point at the same spot each time. When the inch is broken off, move the chisel along and repeat the process. To cut across a stone, keep moving the chisel along the line of cut as you strike it. Don't expect hard, igneous rock to give way easily. It will take a lot of vigorous pounding.

Stone may be cut or trimmed on the ground (left) or on a table covered with sand to provide a cushion (right).

To remove just the edge of the stone, position the chisel's point at an angle to the edge and tap lightly.

Don't be surprised if the stone doesn't break where you want it to; this is all too common even with experienced masons.

Although few masons do, you should always wear glasses of some kind, preferably safety glasses, when cutting stone. The chips can fly with considerable force. Always wear a glove on the chisel hand for protection if you miss the chisel — which you won't if you keep your eye on it.

Laying stone. Stone is laid so each piece is joined to its neighbors by a layer of mortar. Start by spreading a layer of mortar on the footing. Then place the first stone in the mortar. Butter the end of the second stone and place it alongside the end of the first. This operation is repeated until all the stones comprising the first course are laid. The tops of the first course of stones are then covered with mortar and the second course laid in place, care being taken to see that the vertical joints are not aligned. (For details on footings, see Chapter 17.)

When you are working with irregular stones, fill the spaces between them with small stones and stone chips covered with mortar. In this way you fill what would otherwise be voids and fit stones that don't fit by patching them with mortar and small stones.

To fill open areas behind the stones forming the outsides of the wall, make a

Leveling a stone wall with mortar before adding another course.

mixture of mortar and small stones using a little more water than before and just slosh the mixture into place. If you do not have sufficient small stones and chips, purchase gravel or crushed stone and mix it with sand and cement to fill an opening.

When using large irregular stones, most masons drag a stick between the stones to force the mortar as far back between them as possible so it is hidden. When working with cut stones many masons use a small trowel to produce a raised, flat-topped joint surface, which may be as much as 1/2 inch across and 1/4 inch high.

Top the wall either by giving it a smooth, slightly pitched mortar surface or by laying large, smooth stones on top or by capping it with slate or flagstone. Generally, for the sake of appearance, a stone wall is not capped with concrete.

Foundations and retaining walls. Although most municipalities do not permit the use of cut stone for foundation walls—they term this a rubblestone foundation—there is no reason why you cannot make the foundation and even the walls of a one-story cottage out of this material where there are no local codes. If you make the foundation walls about 2 feet thick and the footings proportionally larger, you will have ample support for any one-story house. Make the walls themselves about 1 1/2 feet thick.

Do the same for a mortared-stone retaining wall; a thickness of 2 feet is more than adequate as long as the height is less than 5 feet. Just don't forget to provide the wall with sufficient drainage (see Chapter 4).

Spaces behind the face stones can be filled with small stones.

Raking the joints in a wall with the end of a stick, to force the mortar as far as possible into the crevices so it is concealed.

One way to top a stone wall is to level it with mortar. Although it cannot be discerned from the photo, this top pitches at an angle of about one-half inch per foot.

Partially mortared walls. If you don't want to go to the trouble and expense of digging a trench and constructing a footing, and are willing to chance a few cracks, you can construct a stone wall following the dimensions suggested in the beginning of the chapter. Use just a little mortar, keeping it back from the outside surfaces of the wall so that it is hidden. Use the mortar mainly to join the small stones. In this way the wall will retain a little flexibility and it won't be severely damaged by frost heave.

A wall can also be topped with large stones. These form an attractive surface and provide added strength.

STONE
RETAINING WALLS

As you may have learned the hard way, it is difficult to start and to keep grass growing on a hillside. A heavy rain will often wash your efforts away. And if you have built a rock garden as a means of holding the soil in place and hiding a raw hillside, you have learned that rock gardens require a tremendous amount of your time if they are not to become weed gardens. In addition, a heavy downpour will still wash a goodly portion of the soil from the hillside.

There are two ways to treat this problem. You can ignore it, in which case the hillside will eventually turn into a gentle slope, and you will have no trouble keeping it covered with grass. Or you can retain the soil with an easy-to-build, dry stone retaining wall. Not only will the wall hold the hill in place, it will eliminate the recently excavated look from that portion of your property, replacing it with the finished air of permanence that only masonry provides.

DESIGNING THE WALL

This is simplicity itself, but you must observe a couple of restraints. One is height and the other is tilt. If you give your retaining wall sufficient tilt you can make it as high as the sky. But if you want to make it perfectly vertical or almost so, you run into building code problems. Most municipalities require that all retaining walls higher than 5 feet be "designed." This means simply that you either have to come up with wall dimensions and materials that satisfy the building department or you have to hire a licensed engineer to design the retaining wall for you. All this may appear foolish, but in the past people put up retaining walls without much thought to safety. When the earth behind such walls became water-soaked and soft, the resultant mud pushed the walls over. A 6-foot wall can be a killer when it collapses.

If you are going to erect a nearly vertical wall, hold its height to 5 feet. But if you tilt the wall, its height can be safely increased. To be sure of both their legality and structural soundness, discuss your plans with the building department before starting the job.

A vertical retaining wall made of carefully selected stones.

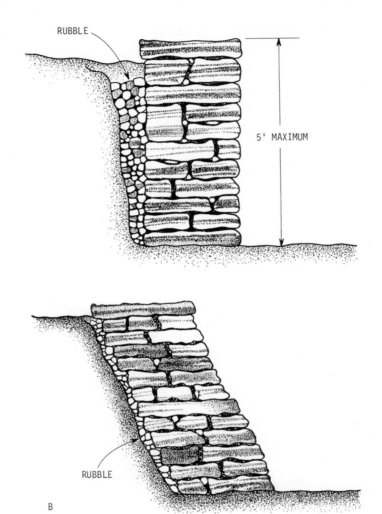

RUBBLE

5' MAXIMUM

RUBBLE

B

Two examples of dry stone retaining walls. Top: A perfectly vertical wall made of flat, trimmed stones. This wall can be 1½ times higher than its thickness, but not higher than 5 feet (4 feet in some localities) without permission from the building department. Below: This wall slopes steeply backwards against the hill. Since most of the weight is against the hill, stones of almost any shape can be used.

In the previous chapter the rules of thumb for vertical walls of various materials were given. The same rules apply to retaining walls if you hold them vertical. However, by tilting the front face of the wall 20 or more degrees you can cut the required wall thickness by a third or more. Tilt the wall 45 degrees and you can get by with just one layer of stone because most of the stone's weight is carried by the hill.

As before, the best kind of stone is whatever you can get for free. If you have to buy, quarry rubble is no doubt the least expensive. In most cases, rubble will be good enough, because when you lean the wall against the hill you can make do with smaller and irregularly shaped stones.

THE ROMAN TECHNIQUE

The method used by many Italian masons for making retaining walls without mortar supposedly originated with the Romans. The technique is fast and easy, and the resulting wall is strong and very durable.

Start by digging out the face of the hill to form an almost horizontal shelf which tilts backward slightly toward the hillside. Do not discard the soil and small stones, but keep them nearby.

Spread the stones you are going to use along the ground 5 or more feet away from the bottom of the cut you have made into the hillside. Stretch a line or lay a long board down to mark the front edge of the wall you plan to erect. You are now ready to start.

As when making an ordinary dry stone wall, use the heaviest stones for the first course, which in this case consists of a single row of stones, 6 or more inches clear of the hillside. Next, fill the space behind the stones with a mixture of soil and small

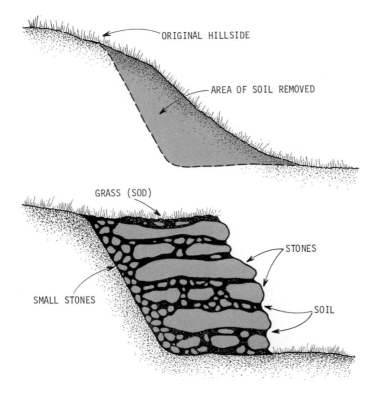

ORIGINAL HILLSIDE

AREA OF SOIL REMOVED

GRASS (SOD)

STONES

SMALL STONES

SOIL

Cross section of a retaining wall made by Roman method.

stones. Tamp this down with your shovel. Then add a few more inches of soil. Cover the tops of the stones, the cracks between them, and the area behind them. With the back of the shovel tamp this layer firm and fairly level.

Now lay the second course of stones so that they overlap the joints between the stones in the first course and so that they are set a few inches farther back than the first course. Place dirt and small stones behind the second-course stones, then cover them with a few inches of soil. Now you can continue upwards the same way until you reach the top of the wall.

CONSTRUCTING A LOW RETAINING WALL BY THE ROMAN METHOD

1 Dig out the hillside to form a horizontal shelf that tilts slightly *toward* the hillside. Use a board or a stringline as a guide to align the front row of stones.

2 Lay a row of small stones along the shelf and cover them with a layer of soil. Place the second row of stones on top of the soil, keeping them level.

3 Cover the second row of stones with a layer of soil and lay the third row in exactly the same manner.

4 Top the wall with large stones or, if you wish, with a "rug" of sod, as shown below.

Finishing the top of the wall. Select your stones, if you can, so that the last course ends about 4 inches below the desired final height of the wall. At this point you have several options open. You can top the wall with the best, largest, and flattest stones you have; you can top it with a grass "rug"; or you can use the small stones you have left to finish the wall and join them with mortar, as previously suggested.

Since the other methods of topping a wall have been covered in the previous chapter, only the rug method will be described here. It is simple enough. Pile sufficient soil on top of the wall to bring it about level with the top of the hill behind, or about 4 inches of soil if that suggestion is impractical. Then cut grass sod into squares and place the squares on top of the soil. Some of the soil will wash out, but the sod will take hold and grow, giving it all a nice rural touch.

STONE STEPS

Like other types of construction described in this section of the book, dry stone steps depend on the weight of the stones and their placement to hold them in place. And like the other dry stone constructions, these steps have an informal, rustic charm that cannot be duplicated by other methods and materials. And further, like all the other constructions covered so far, dry stone steps are flexible and therefore impervious to any movement of the earth, short of a quake.

DESIGNING THE STEPS

There is only one constraint to designing dry masonry steps, and you can even ignore it if you wish. Very simply, you need to know the rise, meaning the vertical distance between the bottom of the staircase and the top, if you wish to divide the distance into a number of equal steps. But you don't have to do so. You can guess at the height of each step as you go along. This will keep the people using your steps on their toes. Normally, we feel for the first step, and since we are accustomed to all the following steps being the same height, we automatically adjust our stride to them. If the rise, as it is called, of each step is different, one has to adjust for every step to avoid tripping.

To find the total height, drive a tall stake vertically into the ground at the foot of the hillside (see page 73). Then stretch a line horizontally from the top of the hill to the stake. Hang a line level on the string to make certain it is level. Measure down from the point where the string touches the stake to the ground. This distance is the total height, or rise. Divide this distance into any number of equal parts, each representing the rise of one step. However, use sufficient steps to hold each rise to 8 inches or less. A higher rise makes for uncomfortable climbing.

Next you need to decide on the width and depth of each step, or tread. Again, the only rule you need follow is to avoid going below the comfortable minimum if you can help it. This is a step width of 24 inches and a depth of 10 inches.

This stairway of large, flat fieldstones was built without mortar, yet it is stable and durable.

Materials. Use flagstone or bluestone, 2 inches thick, for the treads, the stones you step on. Do not use slate, since it is too slippery when wet. You can also use a number of solid concrete blocks, but the results aren't too attractive. And, of course, you can also use fieldstone, if you can find pieces of the correct size and shape.

In addition you will need fairly regular stones for the risers—the stones that go between the treads and establish the height of the steps—and stones for retaining walls to hold back the earth that is exposed when you cut your steps into a hillside.

Precast concrete rounds serve as attractive treads in this stairway. Risers, concealed beneath the curved fronts of the treads, are simply bricks laid on edge.

Having decided on the width and depth of each tread, add 6 inches to the width and 4 inches to the depth. These portions of the tread will be covered by supporting stones and sidewall stones.

To put all this into more specific figures, assume you want the finished tread to be 12 inches deep and 24 inches wide. Adding 4 inches to the depth and 6 to the width brings you to a stone that is 16 by 30 inches in size. Assume further that each tread is going to be 2 inches thick and that each riser is to be 6 inches.

CONSTRUCTION

Let us assume you have secured all the treads necessary, plus a small pile of moderate-sized stones from which you can select stones of the size and shape you will need. The desired tread dimensions are 12 by 24 inches and you have pieces of flagstone cut to 16 by 30 inches. You are ready to go.

Start by cutting into the hillside with a shovel to make a dirt step or shelf 4 inches high, 30 inches wide, and 16 inches deep, with the rear of the shelf slightly higher than the front. Still using the shovel make the front edge of the step or shelf vertical.

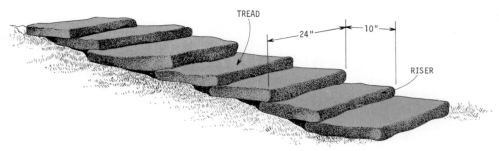

Minimum dimensions of treads for stones steps.

Next, find a number of stones close to 4 inches high. Place them in a row on the ground, tightly up against that 4-inch-high shelf you have just cut. If the stones are a little high, pound them into the soil or remove some soil from beneath them. If they are a bit low, spread some soil on top of them. Now, lay a stone tread down on the shelf. Let the front edge of the tread extend 1 inch or so beyond the row of stones. With a spirit level make certain the back edge of the tread is a bit higher than the front edge. Otherwise rain water will be trapped there. If it is no higher, lift the tread, and sprinkle some soil under it to give it the desired angle.

Next, cut another step into the hillside exactly 4 inches above the top of the first tread. Again make the cut exactly 30 inches wide and 16 inches deep, with the rear of the step or shelf a fraction of an inch higher than the front.

Place a row of 4-inch-high stones in a line across the rear of the first tread and tightly up against the side of the hill. If a stone is more than 4 inches high, you must

CONSTRUCTING DRY STONE STEPS WITH FLAGSTONES AND BRICKS

1 Cut a step into the bottom of the hillside 4 inches high to accommodate the brick risers, and almost as deep as the length of the stone treads. Lay the bricks along the front of the step, using a board as a guide. Drawing below shows how to determine rise of hillside to figure number of steps.

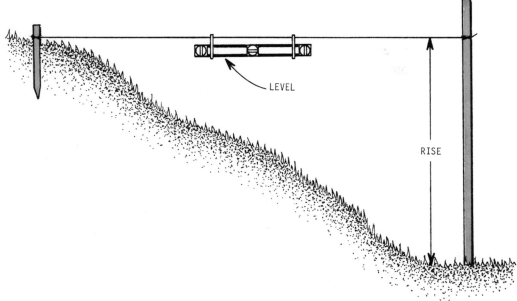

LEVEL

RISE

2 Lay two pieces of 2-inch-thick flagstone in position, covering the
edges of the bricks.

3 Use a level to pitch the flagstones
downward, away from the hill, to
keep rainwater from collecting.

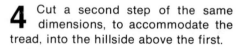

4 Cut a second step of the same dimensions, to accommodate the tread, into the hillside above the first.

5 Lay the flagstone treads and brick risers on the second step. Continue building steps in this fashion until you reach the top of the hill.

remove it. If it is less than 4 inches high, spread some soil on top of it. Now, lay the second tread down and check it for pitch as before.

This procedure is repeated as many times as necessary until you have installed all the steps.

To finish the job, all you need do is construct a small retaining wall on each side of the steps to hold the exposed earth and the stone treads in place. These retaining walls are constructed exactly as discussed in the previous chapter. However, narrower stones are used. In this example, we allowed 3 inches on each side of the stone tread to support the retaining walls. However, you can increase the space on the tread allotted to the retaining walls, and you can also cut deeper into the earth to allow you to use wider stones. The entire width of the retaining wall does not have to rest on the flagstone tread.

VARIATIONS

Should you wish to prevent grass and weeds from growing out from between the stones, you can use mortar to join them. This may lead to some cracking, but generally it won't be too visible because of the way the stones are arranged.

Mix 1 part ordinary cement with 4 parts sand and add sufficient water to make a mush with the consistency of thick sour cream.

When you lay the first row of stones, use a little of the mortar between each stone and then spread a layer of mortar on top of the row. Then, when you lay the tread on top, the mortar will help hold everything together and prevent grass and weeds from sneaking through. Do the same with the following rows of stones, using a little mortar beneath the small stones as well as on top. Use any kind of trowel or flat stick to spread the mortar.

When you construct the small retaining walls, do the same. Spread a little mortar under the first row of stones and use a little mortar between individual stones.

When you have finished, take a round twig about ½ inch in diameter and draw its end over all the mortar that is visible. This will press the mortar more deeply into the cracks and make the surface of the joints neat and uniform. Remove all the mortar that has fallen onto the stones and wash the stones clean.

PART II

WORKING WITH CONCRETE

Concrete is probably one of the most widely used of all building materials, and with good reason. It is strong, permanent, comparatively inexpensive, and easily molded into any of a thousand shapes. It is used for walks, walls, bridges, driveways, and skyscrapers. For a while it was even used for fireboat hulls.

Less generally known is the fact that concrete is far easier to work with than you might imagine from watching a crew pour a concrete bridge or finish a cellar floor. In almost all these operations the concrete is molded—literally poured into a prepared mold and permitted to harden. The molds are almost always made of wood, and therefore masonry work entails some carpentry, but of such a rough kind that carpenters working on wood forms for concrete are usually called wood butchers.

Concrete work is to a surprising degree much like carpentry in that it is repetitive. Once you have learned how to cut a board square with a saw, you can cut all kinds of boards. Once you have learned how to construct a form and fill it properly, there is no limit to the kinds of forms you can make, and no limit to the kinds of concrete shapes you can make, for a walk or a patio or porch is little more than a molded concrete shape.

The pages that follow will tell you just about everything you need to know about working with concrete. If you read all the information carefully before you start, you will have no difficulty constructing walks, drives, patios, slab foundations, piers, and other useful and decorative projects.

TOOLS FOR WORKING WITH CONCRETE

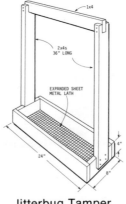

Jitterbug Tamper

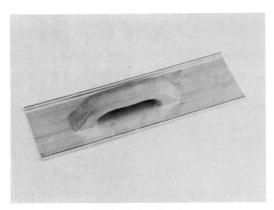

Aluminum Float

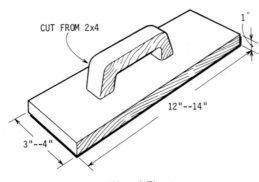

Wood Float

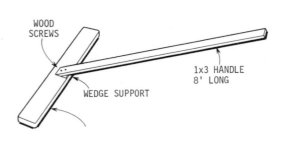

Bull Float

78

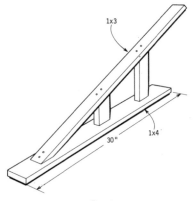

Darby

Finishing Trowel

Pointing Trowel

Edger

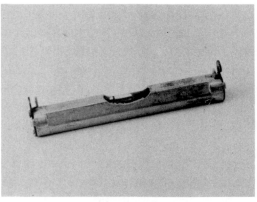

Line Level

ALL ABOUT CONCRETE

You don't need to know all about concrete in order to work with it. Many professional contractors know less about it than the little contained in the following pages. However, you will find working with concrete easier, more satisfying, and much less prone to failure if you read this chapter quickly. Later, should questions arise in your mind, you can refer back. There is no need to memorize this material; just gain a general understanding of the subject.

Concrete is made by mixing sand, water, cement, and small stones together. In a short time, the mixture stiffens and then hardens into a rocklike mass that can withstand the elements for centuries.

Cement is the "glue" that holds the sand and the stones together. The cement used today is called portland cement. It differs principally from the cements used previously in that it will harden under water. Portland cement was invented and patented in 1824 by Josep Aspin, a brick mason of Leeds, England. Aspin made his first batch of cement by heating a mixture of clay and limestone in his oven and then grinding the resultant clinkers to a powder. He named his cement portland because its color resembled that of the stone quarried for building purposes in Portland, England.

Modern portland cement is manufactured in much the same way. Lime, silica, alumina, and other ingredients in smaller quantities are heated in a rotary kiln to about 2,800 degrees F. The materials melt and form clinkers, which are cooled and ground into the powder we call cement.

CEMENT

There are five major types of cement and a number of variations. All types are equal in strength and permanence, and all cements manufactured by all companies are equal in quality, so let price be your guide.

Type 1 is a general-purpose cement.

Type 2 differs from Type 1 in that it has a lower hydration heat, which means it doesn't get as hot shortly after being poured into position. This type is useful for massive structures such as thick retaining walls and bridge abutments.

Type 3 is called high-early because it hardens much more quickly than the other types. It is useful for multilevel, poured-concrete structures where it is important to take down the forms quickly so that work can proceed on the next stage or level.

Type 4 is called low-heat portland cement because it generates even less heat than Type 2. If the project is to construct another Boulder Dam, this is the cement to use.

Type 5 was developed for its resistance to the sulfate action that results from contact with soils and water high in alkali content.

All you need remember about these five types of cement is that Type 1 is the one to ask for, but if it is not available, any of the others will do just as well.

Air-entraining cement. Cement types bearing the letter A in addition to their usual number are air-entraining cements (Type 1A, 2A, 3A). These cements contain additives that cause the cement and the concrete made from it to form and retain billions of microscopic air bubbles. These reduce the quantity of mixing water needed by about half a gallon per bag of cement, improve the workability of the mix, and most important, increase the resistance of the concrete to frost and the effects of ice-melting chemicals such as calcium chloride and common table salt.

For these reasons it is always advisable to use air-entraining cement to make concrete for walks and driveways in frost country. There is little need for this cement in vertical structures because they usually drain fairly well and retain little moisture.

If there is nothing but air-entraining cement to be had, use it exactly as you would any other cement bearing the same type number. On the other hand, if you require an air-entraining cement and there is none on hand you can convert any standard cement to an air-entraining one by adding the necessary chemicals yourself. The only proviso is that the cement and concrete must be machine mixed. You cannot mix the chemicals and cement sufficiently well by hand to make the results worth the effort.

Vinsol resin, Airolon, N-Tair, and Darex are the names of a few of these chemical products. Darex is added at the rate of two tablespoons for every bag of cement mixed and is the additive most likely to be stocked by your local masonry supply yard.

Cements of a different color. Where standard cements, which cure to a dull, gray color are unsatisfactory because of color, you have the option of using white or brown. The white cement is available in Types 1, 2, and 1A. The brown comes in Type 1 only and cures to a warm, buff tone that looks much like adobe.

Except for their difference in color, these cements behave exactly like the gray cements of the same type.

Storage. Cement is packaged and sold in sealed paper bags, each of which contains exactly 1 cubic foot and weighs about 94 pounds. The bag is neither moisture-proof nor moisture-resistant. Cement bags therefore cannot be placed directly on the ground but must always be supported by planks raised in some way a few inches above the earth. This is necessary to ensure the free flow of air beneath the planks. The tops and sides of the bags of cement must be protected from the rain. The simplest and least expensive way to do this is to purchase a plastic sheet from the masonry or lumber supply yard.

Indoors, of course, you need not take these precautions. However, if you plan on storing one or more bags of cement more than a month or two, seal each bag in a

To test an aggregate for silt content, place about 2 inches of the sand or stones in a glass jar. Fill the balance with water and shake gently. If, after settling, there is more than ⅛ inch of silt atop the aggregate, the aggregate must be thoroughly washed before using. *Courtesy Portland Cement Assoc.*

large plastic garbage bag. Otherwise the cement will eventually absorb moisture and harden.

Bags of cement that have been stored awhile often develop a hard surface layer immediately beneath the paper. When you can easily break this layer into powder with your hand, the cement has merely packed down and can be used. If the layer cannot be easily broken, it has absorbed moisture and cannot be used. If the balance of the cement feels perfectly dry you can probably use it. If there is any doubt, however, it should be discarded. Once cement has absorbed water and hardened it is useless as cement even though it may retain some of its original powder form.

AGGREGATES

Concrete is densest and therefore strongest when its aggregates—the sand and stone—consist of particles of many different sizes, because that is when they pack together most closely. However, while the ultimate in concrete strength may be necessary when building a Hoover Dam, it is totally unnecessary when constructing a driveway or a foundation for a house.

For ordinary, everyday constructions, the range of particle sizes found in whatever sand is available is more than sufficient, and the range of sizes in ordinary, casually screened crushed stone or gravel is equally satisfactory. There is no need for "sharp" sand, in other words sand with particles having sharp rather than rounded edges. In fact, tests have shown that while cement adheres with greater firmness to sharp particles, more cement adheres to rounded ones. The end difference in concrete strength is therefore nil, and there is no point in searching for sharp sand when ordinary sand is at hand.

Cleanness tests. The cleanness of the aggregates is much more important than a range of sizes and sharp points in their particles. If you purchase sand and stone from a commercial supplier, you can rest assured that the aggregates are satisfactorily clean. If you are securing your aggregates from a lower cost source such as a sandpit or a convenient gravel bank it is advisable to check for the presence of silt and organic matter, either of which can seriously weaken your concrete.

To test for excessive silt, place about 2 inches of the aggregate in a glass jar. Fill the balance of the jar almost to its top with water. Attach and cover and gently shake the jar until the silt has been loosened. Let the bottle stand until the water clears.

Then measure the thickness of the layer of silt that may have been deposited on top of the aggregate. If there is a layer $1/8$ inch thick or more, the aggregate should not be used without a thorough washing. This is easily done by spreading it over a clean, pitched surface such as a driveway or the tilted back of a dump truck and hosing it down until the runoff is reasonably clean.

To test for organic matter, place approximately half a pint of the aggregate in question into a glass jar. Add an equal amount of clean water and then a heaping teaspoonful of household lye. Cover the jar and shake it gently until the lye has dissolved completely and the solution has thoroughly washed over all the aggregate. Let the bottle stand for twenty-four hours and then inspect its contents in a strong light. If the color is deeper than a very pale orange the aggregate contains too much organic material and must be washed.

Coarse aggregates. Crushed stone and gravel are the most common types. The first is stone that has been mechanically crushed. Generally it is screened so that you can purchase it in a size fairly close to the one you wish. Gravel consists of small stones that have been rounded by the action of water. Sometimes the stones are naturally all of one size, or they may have been screened for size. Bank-run gravel combines a large range of sizes, often including coarse sand.

Well-graded aggregates have particles of various sizes. This is coarse aggregate consisting of stones from ¼ to ½ inch in diameter. *Portland Cement Assoc.*

Selecting the aggregates. Although it might appear that large, rough aggregates produce a concrete stronger than that made with small, smooth, round aggregates, there is nothing to indicate that this is so. Both gravel and crushed stone make equally strong concrete irrespective of the size of the stones. The only advantage to using large stones is in the lesser quantity of sand and cement needed. This saving, however, is more than offset by the additional work entailed in mixing, moving, and placing the large-stone concrete.

As a general rule, use ½-inch gravel where you have to move the concrete more than 10 or 15 feet along a form with a hoe or rake. Use stones up to ¾ inch thick when the poured concrete has to be screeded (made level) and troweled (made smooth), as it would be for a concrete walk or patio. When the concrete is to be dumped into a form, the top of which is not going to be made particularly level or smooth, as for example, in a building footing or foundation, you can use stones up to 2½ inches thick.

In no circumstances should you use stone that exceeds a quarter of the thickness of the poured section. For example, never use stone larger than 1 inch when pouring a 4-inch-thick slab, or larger than 2 inches when pouring an 8-inch-thick wall.

If the form contains reinforcing steel bars, no stone should be more than three quarters as large as the spacing between the bars. This is to make certain no stones get caught between the steel bars and prevent the passage of concrete during the pour.

SETTING AND CURING

Setting. As soon as water is added to cement and aggregate, the mixture begins to go through a series of irreversible physical and chemical changes on its way to becoming the monolithic, rock-hard substance we call concrete.

These changes are not evident while the ingredients are being mixed, but become obvious shortly after the concrete is permitted to rest, either in a form or on the bottom of the mixing pan.

Hydration causes the temperature of the mix to rise. The larger and heavier stones sink a little toward the bottom. Whatever uncombined water is present rises to the surface in a process called bleeding. The concrete slowly loses its plasticity and stiffens. When you have to break it to change its shape, it has reached the point called initial set. Ten or so hours later the concrete is even harder and reaches a condition called final set.

The approximate moment of initial set is important for several reasons. First it indicates the general condition of the mix. With air and material temperature at 70 degrees F, the time lapse to initial set should be about one hour or more. If the concrete sets much more quickly than this, it is very likely that insufficient water was used in the mix. Initial set also limits the time available to finish the surface of the concrete. Generally you can start to trowel or brush concrete shortly after bleeding and can continue, if necessary, for a little time past initial set, but not much longer. The concrete then becomes too hard.

The setting time is shortened by heat, wind, and sunshine. It is increased by cold, moist air, more water in the mix, and the sprinkling of additional water on the surface of the concrete.

Final set marks another important change point in the condition of the concrete. Up until final set the concrete is called green because it is more or less green in color. Until final set you can, if you wish or need to, crush the concrete with a hammer, mix

This graph shows how the strength of placed concrete increases with the passage of time.

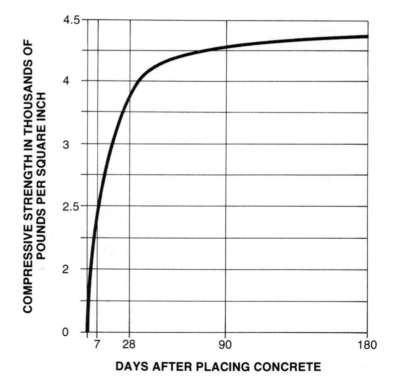

it with water and use it again as fresh concrete. This should be borne in mind. For example, if you find a small hole in your concrete and have nothing on hand to plug it with but a piece of discarded, but still green, concrete, you can break this up and use it for a patch.

Curing (hardening). Mixed with sufficient water and prevented from drying out, concrete grows harder and stronger with the passage of time. The process is called curing and starts sometime after final set and continues for years, perhaps indefinitely. The gain in strength is most rapid at first, tapering off until it becomes almost imperceptible after the first few years. Typically, concrete that exhibits a compressive strength of 1,500 pounds per square inch (psi) three days after pouring resists 2,000 psi after seven days, 4,000 psi after twenty-eight days, 5,000 psi after three months and 5,500 psi after one year.

Drying can be prevented by covering the concrete with a waterproof covering, wet newspapers, wet straw, or sawdust or by sprinkling it frequently with water after it has gone well beyond initial set.

Whether or not lack of water in the mix can be corrected after the concrete has reached initial set is a moot question. However, no harm can be done to set concrete by hosing it down, so it is worth a try.

PROPORTIONING THE INGREDIENTS

The quantities of cement, sand, stones, and water you mix together to make a batch of concrete are important but not critical. There must be sufficient cement to cover every particle of sand and sufficient sand to cover every stone and fill all the spaces between the stones. There must also be sufficient water for hydration and curing. These conditions are easily met by proportioning the ingredients by formula and adding the water by eye.

The more or less standard formula for watertight concrete subjected to moderate wear, as for example, in floors, walks, drives, swimming pools, etc., is 1 part cement to 2 parts sand and 3 parts stone, by volume, with maximum aggregate size limited to under 1½ inches. In practice, however, this ratio is modified to 1 : 2¼ : 3, for reasons explained below.

The standard formula for footings, walls, and similar structures not subjected to wear and/or water is 1 : 2 : 4, which in practice is modified to be 1 : 2½ : 4, with the aggregates under 1½ inches.

The formula for economical concrete suitable for such purposes as heavy foundations, backing up stone masonry, and filling is 1 : 3 : 5.

The reason for adding the small extra quantities of sand to the first two formulas is bulking. When sand absorbs moisture the particles separate and the sand expands. Moderately damp sand containing as little as 7 percent moisture may bulk up to 25 percent, that is, occupy 25 percent more space than it would when perfectly dry. By adding the small quantities of sand specified above or even a little more you make certain you have sufficient sand and do not need to estimate its water content. However, sand should always be protected from the rain as it can bulk up to 30 percent if the particles are fine and there is sufficient water present.

Water content. The strength of concrete depends to a great extent on the quantity of cement used, but even more on the quantity of water added. Very roughly, if you go from 4 bags per cubic yard of concrete to 7 bags, using 2½-inch stones, you will increase the 28-day strength of concrete from about 2,000 psi to about 5,000 psi.

On the other hand, if you increase the quantity of water you use per sack of cement from 4½ gallons to 10½ gallons you will reduce the 28-day compression strength of your concrete from about 5,000 psi to 1,000 psi.

Though both statements are absolutely true, they have only too commonly led inexperienced concrete makers into difficulties. While a richer mixture (more cement per batch of concrete) will produce stronger concrete, the additional strength is unnecessary.

In theory you can make concrete with as little as 2½ to 3 gallons of water to each bag of cement. In practice, however, you get into serious difficulty, since this quantity of water will not make a workable mix. Without sufficient water to act as a lubricant the mixture becomes harsh, difficult to mix, difficult to move, and impossible to finish. Moreover you may end up with dry spots where no water has penetrated, unmixed areas where the cement paste hasn't penetrated, and concrete that will not cure and grow stronger with time because it doesn't contain sufficient water for the chemical process of curing.

On the other hand, if you use a reasonable quantity of water, that is, about 6 to 7 gallons per bag of cement, you will avoid all these problems. Mixing, moving, and finishing will be easier. At 7 gallons to the bag, 28-day concrete has a compression strength of 3,000 psi, and you don't have to concern yourself with measuring the

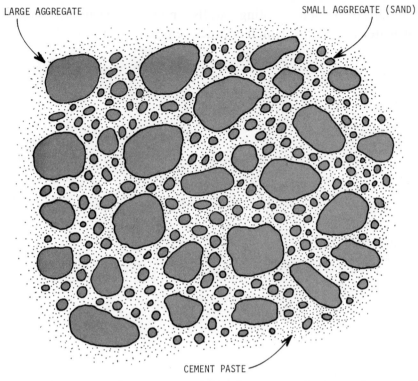

LARGE AGGREGATE   SMALL AGGREGATE (SAND)

CEMENT PASTE

For concrete to harden properly, the cement paste (water and cement) should cover every particle of fine aggregate (the sand) and cement-coated sand particles should contact every piece of large aggregate (stones) and fill every void. There should not be empty spaces nor cement-free areas if the concrete is to achieve maximum strength.

moisture content of the sand because it is easy to judge when you have added 6 to 7 gallons of water to a mix.

Now, to support the statement that there is no need for additional strength beyond that offered by the 7-gallon cement using the formulas given.

Assume you are going to construct a three-story, wood-frame house on a continuous perimeter foundation of concrete block. Assume there are no internal girders nor piers and that the building will be 30 by 40 feet in size and that the foundation wall will consist of 10 courses (rows) of 8-inch blocks.

The total weight of the blocks comes to 38,000 pounds. The house itself and its load (people and objects) comes to about 254,000 pounds. This totals close to

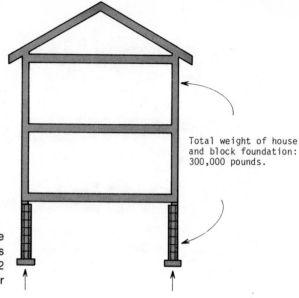

Total weight of house
and block foundation:
300,000 pounds.

Although the total weight of the house is 150 tons, the weight is distributed so that less than 12 pounds per square inch bear on the footing.

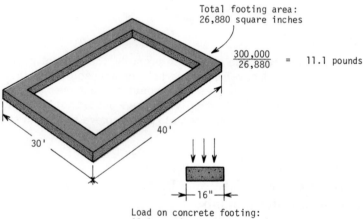

Total footing area:
26,880 square inches

$$\frac{300,000}{26,880} = 11.1 \text{ pounds}$$

40'

30'

16"

Load on concrete footing:
11.1 pounds per square inch.

300,000 pounds or 150 tons, a mighty weight indeed. To support this load you will probably lay down footings at least 16 inches wide, which works out to a total of 26,880 square inches of support. Dividing 300,000 pounds by 26,880 square inches works out to a load of a little more than 11.1 pounds per square inch. Thus you can see that even if you go wild with water, your concrete's *compressive* strength will be many times greater than necessary.

CRACKING

The desire to make concrete as strong as possible no doubt arises from the desire to avoid cracks. Unfortunately, even the strongest concrete will crack when subjected to bending stress. Concrete has a low tensile strength — that is, it has little resistance to pull, which is the force exerted on a beam or bar when it is bent.

At best, concrete's tensile strength, using little water, plenty of cement, and waiting one year for a complete cure, is only 650 psi. Ordinary, 6-gallon concrete develops 500 psi tensile strength in the same time. The 150-psi difference between ordinary, everyday concrete and the superstrength stuff isn't worth the extra cost and effort. This becomes more obvious when you consider that ordinary steel has a tensile strength of 30,000 psi; some steels are rated at over 100,000 psi. This is why steel can be used for a bridge and why a concrete beam will hardly support its own weight without considerable reinforcement.

Cracking can be reduced by the use of reinforcing steel bars and/or steel-wire screen inside the concrete. The word "reduce" is used advisedly, since you cannot eliminate all possibility of cracks by merely using a few steel bars. The reinforcing wire bars and steel-wire screen or mesh, however, are very important where cracking is expected because they hold the parts of the cracked concrete together and so prevent complete rupture.

The only way you can positively eliminate cracks is to make certain the ground upon which you build is perfectly stable and capable of supporting the load. Should the earth settle or move, your concrete will crack unless you have so much steel inside that you have a steel girder rather than concrete. This means that you must dig down to virgin (undisturbed) subsoil; that you cannot build on fill (recently positioned soil); that you should not overload the ground by using undersize footings; and that you should not build where there is considerable ground-water movement, which may shift the soil beneath your footings.

ADMIXTURES

Any number of different chemicals may be added to concrete during mixing to change its nature in some way. One admixture has already been mentioned. It is the group of chemicals used to cause the concrete to entrain air.

Other chemicals are used to slow or speed up the setting time, lower the freezing point of wet concrete and increase the resistance of cured concrete to water penetration.

The chemicals most frequently used are calcium chloride, which increases wet concrete's resistance to freezing, and Anti-Hydros, a trade name for a chemical that makes concrete fairly waterproof. Both speed setting more or less in proportion to the quantities added. Calcium chloride makes the concrete stiffer and more difficult to work, but Anti-Hydros does not.

For more specific information, discuss your requirements with your masonry supply yard.

PROTECTION AGAINST WEATHER

Frost. If wet concrete freezes before it reaches final set, it will revert to a mound of aggregate. Fortunately, it does not freeze easily. Hydration, the chemical action that transforms cement paste into a solid, produces a little heat, and the dissolved chemicals themselves lower the freezing temperature of concrete considerably below that of water.

While all this is true, low temperatures, even if not below freezing, greatly slow the chemical change necessary to final set. This means that you must hold the temperature of the concrete above freezing not for a day or two, but for a week. Otherwise any frost that occurs during the week you are waiting for your unprotected concrete to reach final set can destroy all your work. Generally, a week at 50 degrees F is required as a minimum before the concrete can be safely exposed to freezing. At a lower temperature, an even longer period of time is necessary.

It is therefore foolish and dangerous to pour when there is any chance of freezing weather, especially when the concrete is to be completely outdoors and unprotected, as for example when making a driveway or foundation. In many municipalities the building department will not let you pour when the temperature drops below a safe minimum. Or, they insist on inspecting the job to make certain the proper precautions have been taken.

When your concrete is poured inside or adjoining a building, as in the case of a concrete porch, artificial heat is practical. You can shield the area with a sheet of plastic and heat it with a portable heater, which can be rented.

You can shorten the period during which the concrete has to be protected against frost by using Type 3 cement (high-early) or by adding calcium chloride at a rate not exceeding 2 percent of the weight of the cement used (2 pounds per bag of cement). Any addition in excess of this percentage will seriously reduce the strength of the concrete. The use of Type 3 cement or calcium chloride will speed up setting by about 30 percent; thus the period of protection can be shortened by an equal amount.

Note that calcium chloride should not be used in concrete that is to be reinforced with steel or that will be in direct contact with a steel girder or beam because the chemical will accelerate corrosion.

Note too that cold days can be very dry, and therefore it is important to protect the green concrete from drying out as well as from the frost.

Heat. High temperatures speed up setting, curing, and drying. In hot weather try to use cool or cold water in the mix and add a little more water than normal. Keep the freshly poured concrete protected from the sun and wind and do not delay too long before finishing. In really hot areas, try to do your mixing and pouring early in the morning, before the sun has heated everything up.

PACKAGED MIXES

Dry, premixed concrete can be purchased in paper bags. The mixture is sold under a variety of trade names and is very satisfactory for almost all purposes. However, the price is prohibitive, and if you need any quantity, packaged concrete is not the solution.

On the other hand, when all you need is a little concrete to fill a hole or make a patch, packaged concrete is an ideal answer. It is convenient, and since there is no waste, it is economical for small jobs.

All you need do with premixed concrete is add water. However, since you may not have additional sand or cement on hand to add to the mixture, you must be careful not to add too much water. A cubic foot of dry concrete will require about 1 gallon of water. Add the water very slowly while you mix the ingredients, bearing in mind that as little as a single glass of water can mark the difference between too little and too much. So go slowly after you have added 3 quarts or so.

If you estimate that you won't have sufficient concrete for the job, throw some clean, well-washed stones or bricks into the mix. As long as the concrete surrounds the added material, the final mass will be as strong, or stronger, than it would be without it.

HOW MUCH CONCRETE DO YOU NEED?

It is usually important to know in advance approximately how much concrete you are going to require for a particular job, whether you are buying your concrete already mixed or mixing it yourself. In either case to order too much is a waste of money. At the same time, to order too little and run short in the middle of a job is a great nuisance. It is also costly, since you always pay a higher price for small-quantity deliveries.

Concrete is generally poured into forms, which are retaining walls of wood that have been built to give the finished project the proper shape and dimensions. The concrete is poured into the form in much the same way that you would pour cake batter into a baking tin. The surface of the concrete is leveled with the top of the form by a process called screeding, which consists of dragging and pushing a straight-edged board across the top of the form.

Forms are built for slabs (including driveways and sidewalks) and for walls, footings, and steps. To estimate the concrete necessary to fill a form, multiply its length by its width by its height, all in inches. This gives you the volume of the form in cubic inches. Then divide by 1,728 to get cubic feet, and again by 27 to get cubic yards, because this is the measure commonly used for ordering cement and aggregates.

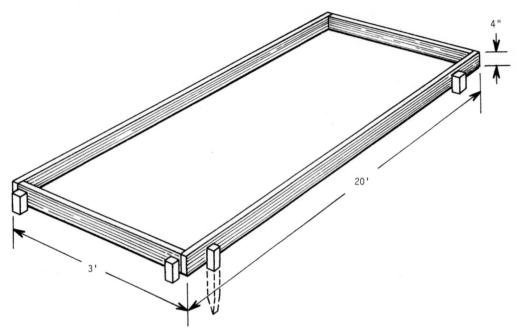

To find the quantity of concrete required to fill a form, multiply the length of the form by its width by its height, all in inches, and divide by 1728.

Example: 20 feet × 12 = 240 inches   240 × 36 × 4 = 34,560 cubic inches
        3 feet × 12 = 36 inches   34,560 ÷ 1728 = 20 cubic feet
(1728 cubic inches equals 1 cubic foot)

Once you have determined how many cubic yards of concrete you need for a particular project, you have to add a certain amount to compensate for waste. Some concrete always falls onto the ground or sticks to tools. Also, the bottom of the form is usually earth or crushed stone and never perfectly flat. Therefore, after you have calculated exactly how much concrete is needed for the form itself you must estimate how much more you need to make up for losses.

If the project is a slab, figure another 20 percent for amounts under 2 cubic yards. Figure 15 percent extra if you are ordering between 2 and 5 cubic yards, and 10 percent extra if you need more than 5 cubic yards.

Wall and footing forms are only variations of slab forms with taller sides. For various reasons, there is less waste when pouring concrete into tall forms, so you can reduce the amount of overage you order. Add 15 percent to a total under 5 cubic yards, and 10 percent when the quantity is over 5 cubic yards.

MIXING YOUR OWN CONCRETE

As you probably know, there are two means of securing concrete for your project. You can purchase the necessary ingredients and mix them together yourself, or you can purchase concrete already mixed and delivered to your door in the form called ready-mix or transit mix. The choice is not always simple, for there are often many factors to consider.

COST

It would appear that you can save money by mixing your own concrete, but this is not always the case. A cubic yard of concrete requires five or six bags of cement plus about a yard and a half of sand and stone together — so that to figure on half a yard of sand and half a yard of stone would be to underestimate by a third or so. Moreover many yards, both lumber and masonry, now charge for delivery and also tack on a healthy extra fee for delivering less than a full load. So unless you have your own transport facilities, be certain to get the full delivered price for all the materials you need before you make your decision as to whether to mix your own concrete or use ready-mix.

Since cement-batching plants may handle thousands of yards daily, they usually purchase their sand and stone by the train load or the barge load, and the local yards frequently purchase these materials from them. So in many instances you will find that a delivered, mixed cubic yard of concrete will cost you less than a yard you mix yourself, even with the penalty ready-mix companies always charge for small loads.

EQUIPMENT

Most localities have one or more "rent-it" shops from which you can rent just about any tool or piece of equipment necessary for masonry and concrete work. If you want to purchase the equipment, possibly to use with your neighbors, look for contractors' and builders' supply companies and write to Goldblatt, 511 Osage, Kansas City, Kansas 66110 for their prices before you turn to local hardware shops.

Very small batches of concrete, say no more than 1 or 2 cubic feet at a time, can be comfortably mixed in a wheelbarrow or a mortar pan. To mix a larger batch at one time without wasting too much effort you need a large, flat, waterproof surface such as a driveway or a concrete floor.

To mix still larger batches at one time efficiently you need a mortar box. Such a box is the most efficient means of mixing concrete by hand. You can comfortably mix up to 6 cubic feet in a small box (53" x 25" x 11"), and up to 15 cubic feet in a large box (82" x 34" x 11").

Small quantities of concrete can be mixed in a wheelbarrow; for larger amounts use a mortar box.

Two power mixers awaiting customers in a rent-all yard. Pipes at the side and rear of machines are parts of scaffolds that are also rented.

The alternative to mixing your own concrete by hand is mixing it with a power mixer. These can be rented with capacities from 1½ cubic feet up. Some of the small models are constructed like a wheelbarrow, which enables you to dump their load wherever you desire. This type can probably be fitted into the rear of a large station wagon. There is also a stand-up portable model with a 2½-cubic-foot capacity that supposedly can fit into the trunk of an ordinary car. Larger capacity mixers have to be towed.

Power mixers eliminate the labor of mixing, but they do nothing else. You still have to load and unload them, and with the exception of the wheelbarrow type, you still have to haul the mixed concrete to where it will be used. In this writer's opinion, unless you have special requirements, there is little to recommend any mixer with a capacity of less than 3 cubic feet over a mortar box. In most instances you are better advised to use the money that you would spend on renting a small mixer to purchase a mortar box, which you can then use for any number of jobs over the years without additional expense.

Using a wheelbarrow to measure and transport crushed stone to a mortar box, where it will be mixed with sand, cement, and water.

A power mixer is no great time-saver if you take into account the time spent transporting it to and from the rental shop. Note also that small-capacity mixers must be completely emptied as soon as a batch of concrete is finished because they cannot be started with any concrete remaining in the drum. Therefore, if you plan to use only a portion of each batch as it is mixed you must provide a mortar box or similar container into which you can dump each load and work from there.

In addition to the major pieces of equipment necessary for home mixing, you will require a mortar hoe and some means for protecting your aggregates and cement from the weather if you cannot store them indoors. You will need a bucket, shovel, rake, and work shoes whether you are working with ready-mix or mixing your own concrete.

CONCRETE MIX FORMULAS

As stated in Chapter 6, three mixtures are commonly used for concrete construction in and around small buildings. They are designated by the ratio of cement, sand, and stone used in each. They are:

1 : 2¼ : 3   The strongest of the three. Used for watertight concrete subjected to moderate wear and exposed to the weather: floors, walks, driveways, swimming pools, etc.

1 : 2¾ : 4   Moderately strong. Used for foundations, walls, and other structures not directly exposed to the weather.

1 : 3 : 5   The least strong of the three. Used for heavy, massive foundations, thick retaining walls, masonry backing, and similar purposes.

None of the mixtures should be used with stones larger than 2½ inches. In all the mixtures it is assumed that the sand will be somewhat moist and therefore in a "bulked" condition—in other words occupying more space than it would if it were perfectly dry. All the mixtures are to be used with the addition of approximately 6 gallons of water to every cubic foot (bag) of cement.

The 1 : 2¼ : 3 mixture is the most expensive because it contains the greatest percentage of cement. It is also the most workable for the same reason. The two following mixtures are proportionately less expensive but harsher and more difficult to work with.

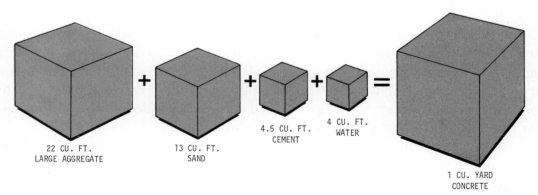

22 CU. FT. LARGE AGGREGATE + 13 CU. FT. SAND + 4.5 CU. FT. CEMENT + 4 CU. FT. WATER = 1 CU. YARD CONCRETE

Amounts of sand, stone, and cement necessary to make 1 cubic yard of 1:3:5 mix concrete.

INGREDIENT QUANTITIES

In Chapter 7 we discussed the steps necessary to estimate the quantity of concrete required for a specific job. We are now going to discuss how to determine the quantities of cement, sand, and stone you need to make a specific quantity of concrete according to any of the three formulas given.

FORMULAS FOR MAKING CONCRETE

To make 1 : 2¼ : 3 concrete

Mix 1 part cement with 2¼ parts sand and 3 parts stone, or
4 parts cement with 9 parts sand and 12 parts stone, by volume.

To make Use	1 cu yd concrete	2 cu yd concrete	3 cu yd concrete
cement	6 bags	12 bags	18 bags
sand	14 cu ft or .51 cu yd	1.02 cu yd	1.53 cu yd
stone	18 cu ft or .66 cu yd	1.32 cu yd	1.98 cu yd
water	36 gals	72 gals	108 gals

To make 1 : 2¾ : 4 concrete

Mix 1 part cement with 2¾ parts sand and 4 parts stone, or
4 parts cement with 11 parts sand and 16 parts stone, by volume.

To make Use	1 cu yd concrete	2 cu yd concrete	3 cu yd concrete
cement	5 bags	10 bags	15 bags
sand	14 cu ft or .51 cu yd	1.02 cu yd	1.53 cu yd
stone	20 cu ft or .74 cu yd	1.48 cu yd	2.22 cu yd
water	30 gals	60 gals	90 gals

To make 1 : 3 : 5 concrete

Mix 1 part cement with 3 parts sand and 5 parts stone, by volume.

To make Use	1 cu yd concrete	2 cu yd concrete	3 cu yd concrete
cement	4.5 bags	9 bags	13.5 bags
sand	13 cu ft or .48 cu yd	.96 cu yd	1.44 cu yd
stone	22 cu ft or .8 cu yd	1.6 cu yd	2.4 cu yd
water	27 gals	54 gals	82 gals

Note that it isn't necessary to measure the ingredients to the exact amounts given. You can round off the quantities. The change will be negligible.

As already stated, it takes about a yard and a half of aggregates plus about 5 cubic feet of cement and some 30 gallons of water to make 1 cubic yard of concrete. The reason for the disappearing act is the nature of the materials themselves. A cubic yard of crushed stone or gravel is not a solid mass. There are spaces between the stones that amount to roughly half the total volume of stone and sand. The spaces surrounding particles of cement amount to somewhat more than half the volume occupied by the cement. Only water occupies its space completely.

Measuring the solids. The most accurate method of measuring the solids is by weight. But since no one apart from a giant cement-batching plant has a suitable

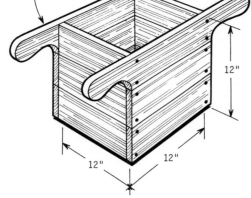

To get fairly accurate measurements, build and use this bottomless measuring box. Just fill the box and lift it up; the contents—1 cubic foot—remain on the ground.

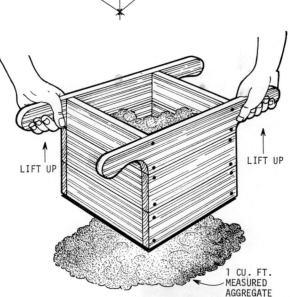

scale, everyone measures by volume. For small quantities of concrete, and when the total amount of concrete is made at one time is unimportant, use a shovel to measure. The method is sufficiently accurate if you don't lose track of the number of shovelfuls you have added and if you take care to scoop up the same amount of gravel as you do sand and cement. Gravel tends to roll off.

For greater accuracy and a foreknowledge of the quantity of concrete you are mixing up, make a bottomless wooden measuring box. Hold the inside dimensions to 12 inches in each direction and you know you have exactly 1 cubic foot when the box is exactly full. Just lift it away and a cubic foot of sand and stone will remain on the ground. Or use a bucket and determine its capacity by dumping a bag of cement inside. Level the cement and mark where the 1-cubic-foot mark is. If a bucket is inconvenient, you can use the same method with a wheelbarrow to mark off 1 or 2 cubic feet.

Measuring the water. Always a use a bucket and keep track of how much water the mix takes so that each batch is not an experiment. Don't use a hose because it is impossible to judge the quantity of water flowing out and because it is often difficult to shut the flow off quickly. The most convenient method of measuring water is to work with an open-top 55-gallon drum. The drum is filled with a hose and you dip your bucket into the drum. If you do not have an open-top drum handy, use a large plastic garbage or trash pail instead.

Use a bucket to measure the water you add to the mix. To save yourself the trouble of opening and closing a faucet each time you need a bucket of water, fill a large plastic garbage can with water and dip the bucket into the can.

ORDERING AND DELIVERY OF THE MATERIALS

Once you have determined just how much of each ingredient you need, add at least 10 percent to allow for waste and error. Generally the yard will send a little more than you order just to be certain, but errors do occur. You can always use any extra aggregate you may receive, whereas a shortage in the middle of a job can be very troublesome.

If your supplier has already mixed the sand and the gravel, and there is no other practical source, ask him about the mixture ratio used and adapt your water and cement to whatever it is. The difference between mixes is not that great. Just don't stint on the cement and the results will be fine.

Much of the labor involved in concrete work results from having to move the aggregate and concrete about. Therefore carefully plan the drop point, which should be as close to your mixing setup as practical, which in turn should be as close to the point of concrete use as possible. If your lot is pitched, the better place to unload the supplies is above the job, rather than below it.

You can drop sand and gravel directly on concrete, but not on blacktop unless you do not mind a semipermanent sandy area. In any case, build some kind of retaining wall—a couple of planks and a few concrete blocks will do—to keep your sand and gravel from spreading all over the place. Then cover the materials with a waterproof cover to protect them from rain, children, and especially dogs.

Build a retaining wall of one kind or another to prevent gravel and sand from spreading over the ground.

If you have no choice but to drop the aggregates on the ground, you should place something underneath to keep them from mixing with the soil. Otherwise there will be some loss, as you cannot use a combination of aggregates and soil. A barrier between the soil and the sand will also help keep the sand dry. A plastic sheet is best. Most masonry and lumberyards stock them. Some contractors use overlapping layers of tar paper, which are better than nothing but rip very easily.

The cement cannot be placed directly on concrete, blacktop, or the ground for any length of time. The bags are not waterproof and will quickly pick up ground dew and moisture. To store cement clear of the ground, raise a couple of planks on concrete blocks, place the bags of cement on top and cover them with a waterproof sheet of any kind.

Be sure to meet the delivery. Don't depend on the driver following your orders. He may well drop the load at his convenience, not yours, and the best a complaint will do is get you an apology.

HAND-MIXING

Taken at a leisurely pace and given the correct equipment, hand-mixing is not at all difficult. Not too many years ago, all concrete was mixed by hand, and it wasn't unusual for men to do this work all day long, which meant at least ten hours.

If mixing is not to be unpleasant and is to be accomplished at a relatively high level of physical efficiency, you have to watch your hands and feet.

Do not wear gloves while mixing. They will slide on the hoe handle, and you will waste energy just holding on. At the same time you must beware of blisters. They can form on soft hands and break before any pain is experienced. To prevent this from happening, keep your hands dry at all times. If you get concrete on your hands, wash and dry them before continuing. Shift hand positions continuously so that the tool handle does not press and slide across the same fold of skin all the time. If your hands get sore, quit or else spread adhesive tape, several layers thick, across the sore skin *before* blisters form. It is too late afterwards.

Wear heavy-soled work boots and heavy socks. You mix concrete with your feet. They provide the base from which you push and pull the hoe. If your feet slide a little, if your shoes do not provide firm support you will waste a lot of energy just trying to get a solid foothold.

Use nothing but clean water for your mix. If you wouldn't drink it, don't use it.

Add about 6 gallons of water for every bag of cement you use. Don't, however, dump the entire quantity you believe necessary into the dry ingredients at one time. Add about three quarters of the water and then go cautiously with the rest. And don't decide you have enough water until the batch is completely mixed. Only then can you be certain.

It is most important to mix thoroughly. The water must surround and dissolve every particle of cement. The cement paste that is formed must be spread over every grain of sand, and every piece of stone must be covered with cement-coated sand. If the cement has not been spread into every minute void, if there are any dry spots, there simply will be no concrete in these places. Remember, the cement is the glue. So mix until your batch is a single, even color with no brown showing, and then mix some more to be certain. While thus engaged, you may attract a number of advisers who may inform you that the more you mix the stronger the concrete will be. This isn't true, but if you do not mix sufficiently, you certainly will have very weak concrete.

Testing. There are two simple tests that will quickly enable you to evaluate the mix you are working on. After you have made the tests a few times, you will never need them again because you will be able to judge your concrete by the way it appears and handles under your hoe.

One is a slump test. Fill an empty 2-pound coffee can (punch a few holes in the bottom) with the concrete to be tested. Invert the can onto a flat surface and then quickly raise it. If the cylinder of concrete that appears retains its form and slumps

The slump test. Make a few small holes in the bottom of a 2-pound coffee can. Fill the can with concrete, turn it over on a flat surface, and lift it clear. The amount the concrete slumps is an indication of its water content. The sample shown has slumped only an inch or so, which indicates the mixture is too stiff and can use a little more water. Concrete that slumps about half the height of the can is about right.

THE MIXTURE TEST

This sample has too high a percentage of stones. Note how they protrude above the level of the concrete.

This sample has more sand than is necessary for normal concrete.

This sample has approximately the correct proportions of ingredients.

less than an inch or so—the can is 6½ inches high—the mix is too stiff and needs more water. (Add it cautiously, it doesn't take much to loosen the mix.) If the concrete slumps to about half the height of the can, the mix is about right for most purposes. If the concrete just spreads out in a limp circle, you have too much water. Add some cement and sand in the ratio called for by the mix.

The second is a mixture test. Place—do not toss or drop—a half shovelful of concrete on a level surface. Run a trowel over the surface of the concrete, using a light downward pressure. If you can clearly see the individual stones, you do not have sufficient cement and sand. If you cannot see any stones at all, you may have too much cement and sand, which is expensive, but perfectly satisfactory from a strength point of view—and also easier to work with. If you can just barely see the stones, and if they disappear just below the surface of the concrete when you push downward lightly a second time with the trowel, the mix is correct for the usual applications.

Mixing in a mortar box. Place the box on a reasonably level, smooth surface. Wet the inside with water. Spread all the gravel you plan to use at this time over three quarters of the bottom of the box. Level the surface of the gravel with your shovel. Now spread all the sand you are going to use at this time in an even layer over the top of the gravel. Follow this with an even layer of all the cement you are going to use. At this point you have a three-layer sandwich consisting of gravel, sand and, cement in the proportions called for by your mix formula.

Now pour a few gallons of water into the end of the box that is not covered by the sandwich. With your hoe slice down through the sandwich and pull some of the material into the water. Work this material into the water, letting the rest of the gravel, sand, and cement be. The batch you are mixing should have much more water than necessary. In fact it should be sloppy wet. This kind of mixture is very easy to work with, since the water acts as a lubricant and vehicle. When this sloppy batch is thoroughly mixed, slice off a little more dry material from the pile with your hoe. Pull this into the batch and mix it in. By correctly judging the quantity of dry material you add you can bring the batch you are working with to the desired consistency. Should you add too much dry material, just add a little more water.

When the partial load is properly mixed with the correct amount of water you can transfer it to your wheelbarrow or directly into a form. Or you can move it to one side of the box and repeat the operation with another portion of the remaining dry material.

Note that since your sandwich contains the correctly proportioned mixture, you

The easy way to open a bag of cement is to place it in the mortar box and cut it with the edge of a shovel.

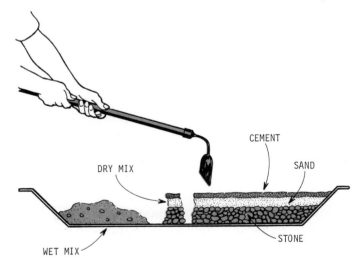

How concrete should be mixed in a mortar box. Since the materials are in layers, if you slice off a section at a time the ratio of sand, stone, and cement is constant.

Using a mortar hoe to mix concrete in a mortar pan.

do not have to turn the entire load over several times to make certain of securing an even dispersal of all the ingredients. The submerged stone and sand weigh less and are therefore easier to mix than the same ingredients dry. With this arrangement you can stop your mixing any time, remove the portion already mixed and continue with the rest at your convenience — providing it doesn't rain in the interim.

Mixing in a mortar pan. You can't use the system mentioned above because you don't have sufficient space. Spread your sand on the bottom of the pan, followed by the stone and then the cement. Don't try to mix more than 1 or 2 cubic feet at a time or you will find yourself severely cramped for space, and the mixing will become very difficult. If you use a mortar hoe, hold it short. If you use a shovel, don't lift the material up into the air any higher than is necessary to clear your shovel when you turn it sideways and drop the material into the pile.

Mixing in a wheelbarrow. Follow the procedure mentioned above. The wheelbarrow is preferable to the mortar pan because you can haul your mixed concrete to the job without the double handling that is involved when you work with a mortar box out of shovel's reach of the form. Use either a hoe or a shovel with the wheelbarrow. To reduce hauling weight, move your wheelbarrow close to where you plan to dump the load before you add water and do your mixing.

Mixing on a flat surface. Wet the surface down to simplify later cleanup. Spread your material in a three-layer circle with gravel on the bottom, sand next, and cement on top. Keep each layer level and smooth. If you are working on a small board, hold the circle to within 6 inches or more of the board's edges. If you are working on a concrete floor or driveway, hold the circle to a maximum diameter of 8 feet if you have a hoe, 3 feet if you have only a shovel. Try to keep the three layers together under 6 inches in height; a greater height makes for difficult mixing.

Starting anywhere a foot or so inside the edge of the circle, pull the material aside sufficiently to add a few gallons of water. Mix enough material with the water to secure a mixture of the proper consistency. Then add more water and mix in some more material, or remove the small mixed batch for immediate use and mix the rest up later. As you remove material from the inside of the circle you can push the material at the circle's perimeter toward the center. As long as you move a complete vertical slice of the dry material into the mixture, it will retain the original ratio.

When you have finished mixing the entire batch of concrete and have removed it

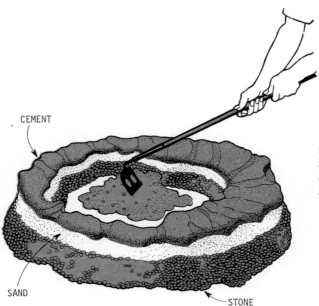

CEMENT

SAND

STONE

Mixing concrete on a flat surface. Spread the material in layers to form a circle, then add a little water to the center of the pile and mix.

all, hose the surface down before you start another batch. This will reduce the possibility of your floor or driveway ending up covered with a layer of cement at the end of the day. If you are mixing in a cellar and the wash water naturally runs down into a drain, it is best to wipe the floor as clean as you can with a rag before hosing. You don't want cement to enter your sewer system. If you are hosing a driveway, there is little harm in the runoff going onto the grass.

MACHINE-MIXING

Selecting a machine. A suitable machine is one that can supply your needs with a reasonable number of batches. If you require a yard, a 1-cubic-foot capacity mixer will have to be loaded and unloaded twenty-seven times. You can mix the yard in less time in a mortar box. On the other hand do not select a machine so large that you cannot handle its output in less than thirty minutes or so. Remember that a cubic foot of concrete weighs 150 pounds, and that more than that in a wheelbarrow is a considerable load for almost everybody.

If you are going to work alone, and if you are going to have to move your concrete some distance and then pour and spread it, as you would if making a walk, for example, you are not going to keep up with a good-sized power mixer. Very likely you will end up with a portion of the mix churning over for a considerable length of time in the machine. Churning, of course, helps keep the concrete from setting, but you cannot extend this period indefinitely. Sooner or later the concrete will set even while it is in motion. If you have mixed more than you can handle you are going to waste the quantity that is left waiting; you have eventually to dump it or it will harden in the mixer. Of course, you do not have to operate a machine to its full capacity, but in this case there is no point in paying the higher rental fee for a large machine.

So give more than a moment's thought to how fast you and your crew, if you are going to have a crew, can handle the output before you decide on the size of the mixer you are going to rent. Mixer capacity is approximately 60 percent of its drum capacity. Note also that single-cylinder engines are very temperamental. Try starting the engine on the mixer you select before you haul it away from the rental shop.

Operating. Set your machine up in a location convenient to both your raw materials and the job. Make certain the machine is level and secure. Use blocks and boards or whatever you have to keep it from "walking" away as it vibrates.

A power mixer in use on a concrete job. Note the mortar box into which the concrete is dumped after mixing.

If you have a gas-operated machine, start it and let it run for five minutes or more to warm it up. Start by dumping a little of the water inside, followed by a little gravel and sand. Add more water, more gravel and sand, and then the cement. You start with water to wet the inside of the drum and lubricate the ingredients so that the machine can mix and tumble them more easily. You add the cement last because it is a powder, and the least amount will be thrown out by the moving paddles when there is a maximum quantity of aggregate to contact.

Be careful when adding the ingredients. Don't let the shovel get inside the mixer. If the shovel is caught by a paddle it will be ripped out of your hand, and if it strikes you, you may never finish the job. Don't look inside while the drum is spinning or the paddles are churning; these machines frequently fling out globs of cement. To get a sample of the mix, pour a little out.

If you are not ready to dump the entire mix and want to let some remain inside, do not stop the machine. Most mixers cannot be started with a load in the mixing drum. Instead let the machine turn over and add some water to the mix to retard setting.

If you go directly to a following batch, just repeat the operation. If there is going to be a thirty-minute or so wait on a warm day, rinse out the drum before you shut the machine down. To do this, toss some water into the drum with the machine in operation, wait a few minutes and then empty the drum before you stop the engine.

At the end of the work day, rinse the inside of the drum, add more water and a bucket or two of clean, dry sand. Then operate the machine for five minutes or more before you dump the sand and shut down. The clean sand scours the inside of the drum and the paddles. Then hose the entire unit down because the cement dust that inevitably flies each time you put cement into the drum collects on the outside of the machine, and if you do not wash it off, it soon forms a thick crust that can only be removed by hammering.

WORKING
WITH
READY-MIX

Ready-mix or transit-mix is the name given to wet concrete carried to a job by a huge truck in a tilted, rotating drum on the truck's rear. When it is certain the truck can drop its load immediately upon arrival, the drum is filled with sand, stone, cement, and water at the plant. The drum is rotated while the truck is in motion, and the concrete is completely mixed by the time the truck reaches the job. When there may be a delay at the job, or when the exact quantity of water to be added is not known in advance, the ingredients are kept dry until the job site is reached. The driver then adds water to the mix as required, taking it from a huge drum positioned above and behind the cab. Mixing is accomplished in a few minutes, whereupon the drum's rotation is reversed and the concrete comes sliding down the chute under the control of the driver, who has stationed himself at the rear of the truck.

Ready-mix is the simplest way of securing wet concrete, and as explained previously, it is very often the most economical way.

In addition to affording an economical price and convenience, doing business with the local ready-mix plant often makes their experience available to you without cost. They know the local building codes pertaining to concrete mixes — which can and which cannot be used for certain work. They can safely recommend a leaner mix (less cement) than you might feel confident in using because they have taken part in thousands of similar jobs. They can also add a variety of chemicals to produce air-entraining concrete, fast-setting, slow-setting, or especially waterproof concrete as

A ready-mix truck delivering concrete. Driver controls flow by opening and closing door at rear of drum.
Concrete already in the chute cannot
be stopped, so you have to signal the
driver before your form is filled.

you may require much more easily than you can yourself. They have the additives
on hand and simply add them to the mix, whereas if you are mixing your concrete by
hand in small batches you have not only to add the special chemicals to each batch,
you have to measure very accurately to make certain all the batches are alike.

116

PROBLEMS

The use of ready-mix poses two problems that can be insurmountable and must be considered before placing your order. They are access and handling. Very simply, can the big truck back up to your job site, and can you handle the quantity of concrete that is to be delivered?

Access. When your job site is close to a good road, the ready-mix truck can back up and deliver the wet concrete by means of its chute without any problem. When the truck has to leave the road, cross a lawn or driveway, pass between buildings or beneath wires, you may have an insoluble problem, which precludes the use of ready-mix.

Mixer trucks are high and wide. The cement company will give you the overall dimensions of their trucks. If there is any doubt as to clearance, measure the space. All you need is a few inches of clearance overhead, but you need several feet of space on either side of the truck, especially if it is going to be close to a building. The reason is weight.

Mixer trucks are very heavy. The big ones can weigh 30 tons and more, plus 2 tons for every yard of concrete they are carrying. The soil next to a building is almost always fill, which means it is soft and hasn't been compacted with the passage of time. If the truck is close and the soil gives way just a little, the upper side of the truck may strike the building. I don't know whether or not home insurance covers damage done by tilting trucks, but few if any mixer companies will let their trucks move over private property without a signed release from the property owner.

Mixer trucks can also chew up lawns, crack any underlying sewer and drainpipes and damage walks and driveways. Whether or not this happens depends mainly on the condition of the soil. If there has been a lot of rain recently and the ground is soggy, it is best to postpone delivery a few weeks. If your plot has been subjected to drought, the chances are that the hard state of the ground will minimize the risk of damage that may be caused by the passage of the heavy truck.

Do not consider laying down ordinary 2-inch-thick building planks as a means of preventing the truck wheels from sinking in. If the earth is soft enough to let the truck tires sink in, the planks will probably be crushed. The only certain support on soft ground is a solid layer of 12x12s, which is of course impractical.

Your best insurance is to ask the cement company to send a man out to inspect the site. He won't take responsibility for the results, but he has or should have suf-

If you have to wheel concrete any distance, construct a road plank. When necessary, butt plank ends together and hold them in position by nailing strips of wood to either side of the planks. Level the planks with blocks of wood.

ficient experience to properly evaluate the load-bearing capability of your soil. And he can also double-check on clearance.

We haven't mentioned open fields because there are no underground pipes in an empty building lot and because whatever troughs the truck's wheels may make in the soil will be covered when you landscape. The mixer trucks have sufficient power to cross moderately soft soil without getting stuck and to go up and down any hill you can walk without holding on to some support.

Handling. When the truck's chute cannot reach your forms you can extend the chute, or use a wheelbarrow or a concrete pump. All ready-mix trucks are equipped with a chute. Some of the chutes are short, while others are more than 10 feet long. Some companies will lend you an extension chute without charge, some charge, and others don't want to be bothered. If you can't borrow or rent a chute, you can make one easily enough. Use a pair of 2x12 planks fastened side by side with cleats, and a pair of 1x8s for the sides as shown in the accompanying drawing. Be certain to provide adequate support as a chute this large, with a 16- x 18-inch cross section, can hold 133 pounds of concrete per linear foot. Remember, too, if there is any drop between the truck's chute and the start of the second chute there is going to be considerable force exerted downward here, so provide generous support and brace the support and chute so that neither can move.

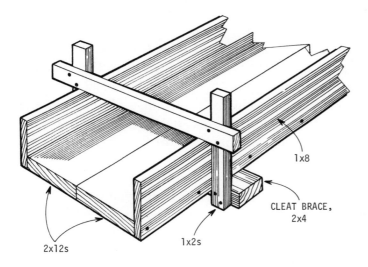

1x8

CLEAT BRACE, 2x4

2x12s

1x2s

A homemade concrete chute and one method of supporting it. The chute should be nailed to the horses when installed at a steep angle.

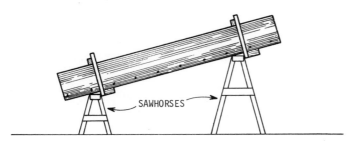

SAWHORSES

To hand-haul concrete you need a mason's wheelbarrow. The usual garden wheelbarrow isn't strong enough. To make it easier on yourself and your helpers, grease the single wheel's axle and make certain that the tire, if it isn't solid rubber, is inflated to maximum pressure. Do not attempt to roll your loaded wheelbarrow over soft earth or rough ground. Make a plank path using either strips of tin to fasten the planks end to end or two strips of wood to either side of the planks. Do not overload the wheelbarrow. A single cubic foot of concrete — 150 pounds — is plenty for the

119

average man, especially when there is some uphill pushing to do. If you overload it is easy to lose control and let the wheelbarrow tip over.

To estimate just how long it is going to take you and your crew to unload the ready-mix truck and move the concrete from the dump point to the form, time yourself walking slowly pushing a wheelbarrow. Figure 1 cubic foot of concrete per trip.

When you cannot reach the forms to be filled with concrete with the aid of a chute, and when it is difficult or impractical to use a wheelbarrow, consider using a concrete pumper to move the mix for you. This is a truck with a hopper mounted on it that is connected to a gasoline-powered pump. The pump feeds a hose. The ready-mix truck dumps its concrete into the hopper; the pump propels the concrete smoothly down the length of the hose to emerge at its end. The hose end is easy to handle, and the thick stream of concrete can readily be guided into a form or several forms. Pump delivery is especially good for deep forms such as walls because the shock of dropping concrete is eliminated and the chance of the form being damaged or displaced is reduced.

Generally, a concrete pumper can drive concrete as much as 500 feet through a hose that is perfectly horizontal, a shorter distance through a pipe that rises 100 feet into the air. Any type of concrete can be pumped as long as the largest aggregate is within the pumper's scope. Generally the limit in stone size is $\frac{1}{2}$ inch, but some pumpers cannot accept stones larger than $\frac{3}{8}$ inch. The mix does not have to be altered otherwise. In fact better results are obtained if no additional water is used.

Pumper capacity is related to hose diameter, which in turn is related to cost. Generally, a pumper feeding a 2-inch hose can deliver 20 yards of concrete per hour. This is more than adequate for the average patio, driveway, house foundation, or wall.

You can locate concrete pumpers by looking in the phone book under concrete pumping service or by asking your ready-mix supplier. Pump rates are not standardized. You have to hire the operator as well as his pump, and some charge by the yard pumped, some by the yard and time; some add a mileage charge to and from the job site. All have a minimum, and they all require some place where they can hose down their pump. So make certain you have the total price before you close the deal and discuss cleaning the pump at the same time.

A pumper is, of course, a tremendous aid when you have to go up stairs or bring concrete down into the far reaches of a cellar, or go through alleys and the like. But pumpers introduce another complication, in addition to cost: you must get both the pumper and the ready-mix truck to the job site at the same time.

As stated previously, you will have no difficulty guiding the open hose end from area to area and form to form. If you are filling forms and nothing more, you can probably do the entire job yourself. However, if you are doing a floor, a driveway, or some similar surface where the concrete has to be spread, screeded (made level), and troweled, don't overlook the help you are going to need for this portion of the job. Finishing labor is not reduced by pumping.

READY-MIX ALTERNATIVES

At present there are two alternatives to ready-mix, not counting mixing your own concrete by hand or machine on the job. The two alternatives are given any number of different names, but essentially the choice is between premixed and do-it-yourself ready-mix.

The first consists of a trailer fitted with a special tank. This is filled with wet concrete containing additives that slow setting time and reduce separation or settling. When wet concrete is shaken, as it is when hauled this way, the larger stones sink to the bottom, and the "fines" rise to the top. This is undesirable as it weakens the concrete. You haul the wet concrete home and dump it where you need it, taking care not to disconnect the trailer for any reason before the tank is empty.

This can be an ideal solution to a job requiring ¼ to 1 yard of concrete when you can readily back the trailer into position for easy dumping. When you have to go from trailer to wheelbarrow, you don't gain much by hauling wet cement home. Obviously, it isn't the thing to do, either, with a light car.

The second alternative to commercially delivered ready-mix is a little ready-mix machine mounted on a trailer that you hitch to your car and drag home after its drum has been filled with the usual mixture of cement and aggregate. Upon reaching home you start the mixer engine, mix away, and dump just like the "big" guys. Since the tilted drum looks like a short, fat cannon, these are often called cannon mixers.

Some of the rental shops have their cannon mixers mounted on the backs of small trucks. This means that your auto doesn't get the beating it would take hauling 2 tons of concrete (1 cubic yard) plus, possibly, another 2 tons of engine and drum.

One major point to consider before renting either of these machines: using them is always a two-man operation. At the very least you need someone to guide you backing up. And, if you are not going to exhaust yourself, you need someone either to control the flow of concrete out of the machine while you spread it, or vice versa. It should be remembered that when the commercial ready-mix truck backs up, there are always two people present, the driver and yourself.

PREPARING FOR READY-MIX DELIVERY

Call your supplier at least a week in advance if you want to be reasonably certain of delivery at the time of your choice. Early-morning delivery is usually best because it gives you the maximum of work hours before dark. But since everyone wants early delivery, you have to call far in advance. This will also give the supplier time to send a man out to inspect the job site should either of you believe it is necessary.

As already stated, discuss your order and your needs to make certain the mix you specify conforms to local building code requirements and that it is the most economical mix for your purpose.

Double-check your forms to make certain they are sturdy and exactly where you want them. Go over the delivery in your mind. Plan where the truck will back up and what forms will be filled at this time. Decide where you and your crew will stand and what each man will do. Get the wheelbarrows and planks set up and ready, if you are going to "wheel" the concrete. Get your water hose ready so that you can wet down the chutes and forms to reduce adhesion. Clean up the area so that there is no chance of anyone tripping on something and so hurting himself. If you are going to use steel for reinforcement, make sure it is ready and in place.

Get your crew on the job early. Run through the steps that will be required of each member, assuming they have not worked concrete before. Make certain the necessary tools, plus a few extra, are on hand. If you are simply going to fill a deep form and strike it off you need a rake or hoe for each man. If you have to spread the concrete, you will need shovels in addition to rakes and hoes. You will also need a screed for leveling the concrete and trowels. Every man should be equipped with a pair of rubber overshoes. The kind without buckles that come up to the knee are best.

ACCEPTING DELIVERY

The driver of the mixer truck will back his truck up to whatever spot you ask him to provided he can safely reach it. He will add as much or as little water to the mix as you specify. If you have a direct drop into a form, ask him to keep it dry. If you have to move the concrete down any distance along a series of chutes, ask for a medium mix, and if you have to move the concrete a long distance along level forms ask for a wet mix. If you are uncertain of what is best, tell the man to hold the water down and give you a sample. From the way the sample handles you can judge how much additional water you require. Remember, of course, that water cannot be removed and there is no dry material on the truck that you can add.

The driver will stand at the rear of the truck and release the concrete at your direction. Make certain he can clearly see your motions and that he fully understands them. If mix drivers have a fault it is overenthusiasm. They tend to release more concrete than you ask for. So indicate to the driver to stop well before you actually want him to, bearing in mind the partial yard of concrete that is on the chute after he closes the drum door.

Part of your deal with the supplier is that his truck remains on the job for a specified length of time—usually a half hour—without charge. Therefore, ascertain the exact length of time allowed you and the rate for extra time. Clock the truck, so that if the driver becomes impatient and starts muttering about overtime, you will know exactly where you stand. Don't let him drop the load faster than you and your crew can accept it. Make certain all your forms are fully filled, even a little overfilled. If the forms are deep, as they will be for a wall or similar structure, poke a board or pole down into the mix every few feet to make certain no air pockets have formed and the concrete has spread into every corner.

When your forms are filled have the driver dump whatever concrete remains in the drum in some convenient spot. If you have calculated your needs correctly, you will have a partial yard left over. This will ensure that you have some wet mix on hand if you have to fill in areas that slump because the form has spread or areas that weren't given sufficient concrete because of the dim light or the rush. If the leftover concrete is not used, it is not difficult to remove if you break it up into pieces just after it sets up a little.

Caution. Moving wet concrete can be physically hazardous. In the excitement of working with a group of men, in the rush to get the concrete into place before it hardens, it is only too easy to strain yourself disastrously. So work cautiously, carefully, and slowly. Don't worry about keeping the water in the mix to a minimum and concrete strength to a maximum. No matter how much water the driver adds to the mix the concrete will still outlast both of you. Wet concrete is extremely heavy and sluggish. If you get desperate, ask the driver to help you; many will do so without being asked. Don't be proud; let them.

10

BORDERS AND LOW WALLS

One fairly simple and inexpensive way to make your grounds appear as though they have regular, professional care is to enclose your flower beds, vegetable gardens, and walks with borders. Then in the fall, you can dump your grass clippings on the flower bed as a mulch, and the clippings will be neatly retained. A border along a flagged path or patio helps keep the flags in place and provides an attractive edge that can hide many minor defects. Along a walk, a border helps keep the grass from straggling over and so saves you the task of edging it. Along a driveway, where a border becomes a curb, it can keep cars from causing damage. So that in a variety of ways borders can produce results far outweighing the modest costs and labor involved.

FLEXIBLE BORDERS

Like flexible flagged patios and stone walls, flexible borders are held in place simply by their own weight and the soil surrounding them.

For materials you can use bricks, concrete blocks, fieldstone, Belgian blocks, quarry stone, and even flagstones. You can position the bricks and blocks vertically,

Flexible borders made of selected fieldstones add charm and beauty to informal gardens.

at an angle, on their sides, spaced apart, or without spaces. You can do the same with fieldstone and cut stone when their shape permits. Borders of flagstone and bluestone don't look very attractive if more than a few inches of stone shows above ground—at least not to this mason.

You can, if you wish, use mortar between the stones and bricks. This removes the flexibility and leads to cracking, but many people do not find the cracks objectionable.

Start by digging a trench to a depth of about a third of the overall height of the material you are going to use. For example, if you are going to use 1-inch-thick concrete blocks on their sides, you would dig a trench about 2 inches wide and about $2\frac{1}{2}$ inches deep. Since the width of a concrete block is close to 8 inches, the $2\frac{1}{2}$-inch

125

trench depth would accommodate roughly the third of the block that has to be set in the ground to hold it erect over a period of time.

The blocks are then placed on their sides in the trench and loose soil is firmly tamped into the spaces between the sides of the trench and the blocks. This completes the job. Nothing can harm a border of this kind. If a runaway tricycle knocks a block down, you reposition it, replace the soil, and the border is repaired.

To hold the blocks — or whatever other material you are using — in a straight line as you position them, use a long, straight board, or stretch a line as a guide. To hold border height at a desired distance above grade, stretch and hold a line with stakes at the requisite height near the edge of the border. To space the blocks or stones evenly apart as you place them in the trench, use a piece of wood or a finger as a guide.

The first step in making a flexible border is to dig a trench. The depth should be equal to one-third the height of the stones or bricks to be used for the border, and a bit wider. A board can be used to keep the trench straight.

Stretch a line along the trench at the proper height and lay the border material — blocks in this case — in the trench. Align the edges of the blocks with the line.

Making a border of common bricks set on end. Board is used as a spacer.

When laying a border of fieldstones, which are usually uneven is size and shape, you have to use your judgment when aligning them with the guide.

A border of common bricks set on end at an angle. A little sand is used between the bricks to hold them apart.

127

CEMENTED BORDERS

Simply cementing one brick or block to its neighbor leads to cracking. To reduce the possibility of cracking to a minimum the bricks, blocks, or stones must be laid on a concrete foundation. Where there is no frost a foundation about 6 inches deep and a few inches wider than the border itself is usually sufficient. Where frost is a problem, the foundation must go below the frost line to be effective and becomes a deep footing.

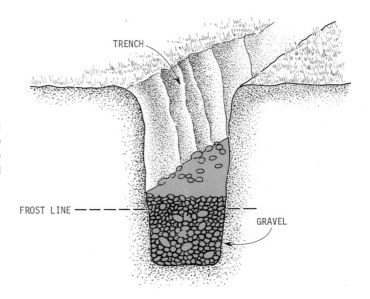

Where the frost line is deep and you want to save on labor and the cost of concrete, dig the trench to a depth below the frost line and then fill it half full of gravel.

Foundations provide advantages in addition to the reduction of cracking, and you may therefore want to set your border on a foundation even though you are not going to cement the component parts together or may even leave a space between them.

Since the foundation presents a level surface on which to work, it is far easier to hold the bricks or blocks in alignment when placing them on a foundation than when burying their ends in the ground. You can also expose a greater portion of the bricks and blocks this way. For example, when you set a brick vertically in the ground, you can expose no more than about 5½ inches above ground. When you set it on a concrete foundation you can expose about 7¼ inches. It is impractical to set a brick on its side in the ground, since most of it is out of sight. However, if you lay it on its

MAKING AN ABOVE-GROUND FOUNDATION AND BORDER

1 Dig a trench for the concrete and construct a form directly above the trench. In this example, 1x3s are nailed to the sides of 2x4 stakes driven into the ground at the ends of the trench. The stakes (only one visible) serve as stop boards and hold both sides of the form in position.

2 The form and the trench below are filled with concrete.

3 The concrete is screeded by sliding a small board along top of form.

side on concrete almost all of it will be visible. Thus, where you need about fifteen bricks positioned vertically to make 3 linear feet of border using the earth alone, by using a concrete base you can lay the bricks end to end and cover the same distance with only four-and-a-half bricks.

Shallow foundations. Excavate a trench the length of the desired border. Make the trench about 6 inches deep and about 2 inches wider than your border material, but no less than 4 inches no matter what the thickness of the material may be. Mix up a slopping wet batch of 1 : 2¾ : 4 concrete. Gently pour the mixture into the trench, bringing the surface almost to grade. Since the mix is very wet it will find its own level. If the ground slopes gently in line with your border, use a little less water in the mix and a float to make the surface of the concrete follow the grade. If the slope is fairly steep, dig a series of stepped trenches, so that your border consists of a number of horizontal sections, on a slightly different level.

Next, well before the concrete has begun setting, place bricks or blocks in position on the wet concrete, pressing each of them in gently a fraction of an inch. Use a

4 Bricks are stood on end in the concrete, spaced evenly with a board. Any suitable type of border material can be positioned this way.

straight board or a line as a guide if you have trouble lining up the bricks by eye alone.

If you wish to join one brick or block to another with mortar (cement) apply the mortar to the side of each brick or block before you place it on the wet concrete. (See Chapter 28 for more information on laying brick.) If the concrete foundation becomes too hard for reasons beyond your control, spread a half-inch layer of mortar over the concrete and set your bricks and blocks in that.

Deep footings. Where the soil is firm and dry, dig the trench at least 6 inches wide and deep enough so that its bottom lies below the frost line. Then carefully and gently pour concrete directly into the trench as before.

Where the soil is too soft to do this easily, dig the trench 8 inches wide and place plywood sheets coated with fresh or used motor oil vertically in the trench, one sheet facing another. Don't bother to cut the plywood to size, but simply let any extra portions of the sheet project above the ground or overlap within the trench. Next, throw stones into the trench between the sheets of plywood to hold their bottoms apart.

131

Have somebody hold the top edges of the plywood sheets apart while you pour the concrete between them. If you have had to dig the trench in the form of a "V" because the soil was so soft, place rocks behind the sheets to keep them close to vertical.

When you have filled the plywood-lined trench with the concrete to grade level or even an inch or so higher, quit. Wait about one day for the concrete to set up fairly well and then gently remove the sheets of plywood that constituted the form. If you have generously coated the wood with motor oil and don't wait any longer than necessary, you will have little trouble removing the plywood. If you haven't oiled the plywood and wait two days or more, you may have a difficult time getting it out.

If you are a cautious gambler, fill the deep trench about one-half full of crushed stone or gravel, and then pour the concrete. The granular support will provide fair drainage and should reduce the chance of frost heave to a great degree. If you are an all-out gambler, just make the foundation 10 or 12 inches deep and about 8 inches wide. By the time spring comes the grass will have grown in and whatever cracks that were opened by frost will be hardly noticeable.

POURED CONCRETE BORDERS AND CURBS

A concrete border or curb consists of the concrete foundation already discussed, but constructed so that its top projects above ground to the height desired.

To protect your border against frost heave with any degree of certainty, you have, as previously stated, to bring the bottom of the border below the frost line. If you want to be half safe, fill the trench half full of crushed stone or gravel, and if you are carefree, just dig the trench deep enough to hold the border in place.

Make the concrete at least 6 inches thick if you are going to make the overall height of the border 2 feet or less, and about 8 inches thick for an overall height of 3 feet. Whatever the height, keep about a third of the border underground so that it cannot be easily knocked down.

Start by digging the necessary trench. If the soil is dry and firm, an above-ground form is all you will need. If the earth is soft and/or wet, you will need a form that extends below ground.

Above-ground forms. To save yourself the trouble of cutting boards lengthwise, make the height of your form equal to any suitable standard board width. Position one board vertically on each side of the trench with the aid of stakes driven into the ground along the outside of the form. Place a spirit level on the top edge of the boards to make certain they are level, then nail the boards to the stakes.

The form—that is, the pair of parallel boards—does not have to be in exact alignment with the trench below. It can be a bit wider or narrower than the width of the trench and can even be off to one side an inch or two.

To construct a form higher than the width of a single board, place a second or more boards above the first. To make the form longer than a single board, position additional boards end to end. Nail a short piece of wood across the ends to hold them together and in line. To make a curved border, bend plywood to the shape you wish.

When you have constructed and positioned the form, close its ends with pieces of wood nailed in place. You are now ready to mix the concrete and pour the form.

Below-grade forms. As previously explained, when the earth is so soft and wet that you can't cut a straight-sided trench, you have to provide a form that extends below grade. Otherwise you will waste a lot of concrete filling the V-shaped trench, and the concrete will be weakened by dirt falling into it.

Start by digging this trench a few inches wider than the desired border width. Cut plywood sheets lengthwise to a width equal to the desired overall height of the border. Place a pair of these cut sheets vertically in the trench, facing one another. The width of the plywood must be such that their top edges are exactly where you want the top of your border to be. For example, assuming you want an 8-inch-high concrete border, and your trench is 24 inches deep, you need sheets of plywood 32 inches wide.

Slop the inner surface of the plywood with motor oil, new or used. Then throw some rocks or crushed stone into the bottom of the trench to hold the bottom edges of the plywood sheets apart. Next, drive some stakes into the ground behind the plywood sheets. Position the stakes to hold the plywood sheets vertical, parallel with one another and with the distance between them corresponding exactly to the desired width of your border. To make a border longer than a single sheet, use several sheets positioned end to end and nailed together with a strip of wood. To save on lumber when making a long border, make it in sections, using the same form several times over.

Next you have to adjust the height of the plywood above the ground and make certain the top edges are horizontal. Do this by either driving the plywood into the ground a little or pulling it up. When done, nail the plywood to the stakes. Finally, nail boards across the ends of the form to close it off.

Use 1 : 2¾ : 4 concrete mix, making it almost slopping wet. Pour it carefully

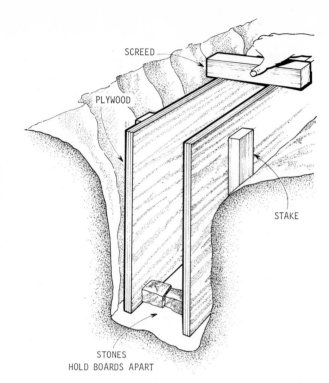

SCREED

PLYWOOD

STAKE

STONES
HOLD BOARDS APART

A simple form can be installed when the soil is too soft to dig a straight-sided trench. Stop boards will be nailed to the ends of the form.

into the form until the form is full. Then poke a pipe or a broom handle gently into the form to help the concrete settle and make certain all the space is filled. Add a little more concrete afterwards, if necessary. Wait until all the excess water "bleeds" to the top. Use a float or a clean, flat piece of wood to push the water off. Then add more concrete and screed it flush with the top of the form by simply sliding your float or a clean, smooth board across the top of the form.

Wait about one day and then carefully remove the form. Fill the spaces between the concrete and the soil with more soil and the job is done.

Pouring borders on existing concrete. Should you decide you would like to add a border to a concrete drive or patio, you can do so without any difficulty. Start by constructing your form, which, if the border is to be a single, straight wall, will be in the form of a long box. Since you will not be able to drive the stakes into the ground to hold the form in place, this must be done another way. One method is to nail the ends of several short pieces of 2x4 to the form, and then place rocks and concrete blocks on the 2x4s to hold the form in place.

Before you place the form, however, take a hose and carefully wash down the concrete that is to be covered. Then give that area a coat of Weld-Crete or Permaweld-Z. Either of these two chemical compounds will help the concrete you are going to pour to bond with the already existing concrete.

Position the form. Mix up a batch of $1 : 2\frac{1}{4} : 3$ concrete, a little on the dry side — or at least not slopping wet. Carefully shovel the mix into the form. Rap the side of the form with the shovel to help the concrete settle. Then screed the top flush with the form.

Wait at least two days, then gently pry the form apart to avoid putting any pressure on the concrete.

Reinforced borders. To make a border or curb strong enough to hold back a rolling automobile, either prepare for the border when you are pouring the driveway or insert reinforcing rods afterwards.

To prepare for the border, insert a number of 6-inch-long, $\frac{1}{2}$-inch-thick steel

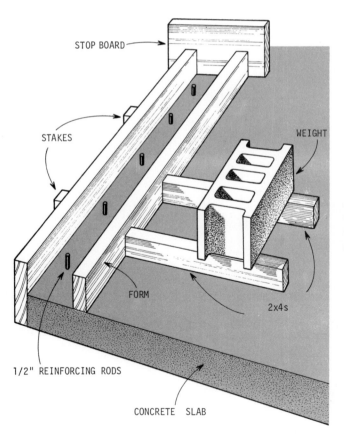

STOP BOARD

STAKES

WEIGHT

FORM

2x4s

1/2" REINFORCING RODS

CONCRETE SLAB

A form for building a low wall or border on an existing concrete slab. Use steel bars only if you expect the border to be struck by an automobile's wheel.

reinforcing bars into the wet concrete, shortly after pouring it. Place the boards about 2 feet part, along the edge that is going to carry the curb. Then pour the curb atop the bars after the drive has set up hard enough for you to work.

To install a reinforced curb on an old driveway, drill $\frac{1}{2}$-inch clearance holes into the concrete about 4 inches in from the edge. Then insert the bars into the holes, position the curb form around the bars and pour as before, using either of the chemicals mentioned to improve the bond between the new and old concrete.

11

CONCRETE WALKS

Beautiful, permanent walks of concrete are easy to construct if you take the time necessary to plan the work properly and if either you have sufficient man power or you break the task up into small sections that you can easily handle yourself, one section at a time.

When you have finished your walk, not only will you have the satisfaction of accomplishment but also that of having pocketed at least three quarters of the sum the contractor asked for the same job.

In the previous chapter we discussed poured concrete borders and low walls. If you recall, the walls were made by constructing narrow forms which were filled with concrete and then screeded level.

Concrete walks are made by constructing a large, shallow form, filling it with concrete and then screeding the surface of the concrete flat.

The height of the form's side determines the thickness of the final concrete slab. The inside width of the form establishes the width of the walk, and its length is equal to the length of the walk. In other words, instead of a long, deep box that we fill with concrete, we construct and fill a large, shallow box that serves as a mold.

Designing a concrete walk involves little beyond a decision on how wide and thick you want it. Walks around a home vary in width from 2 to 5 feet. Anything less than 2 feet wide is difficult to walk on. Sidewalks, which go past the front of the house, are almost always the property and concern of the municipality and are usu-

Two contrasting examples of how a walk can enhance the entrance to a house. The broad walk at right consists of repeating sections of exposed aggregate and plain concrete, separated by wood strips. Below, a simple, narrow concrete strip forms a right angle with the side of the house, breaking the monotony of the broad lawn and providing a tricycle path for small children. *Courtesy Portland Cement Assoc.*

ally 5 feet wide. Walks leading from the sidewalk to the front door are usually 3 to 5 feet wide. Those leading from the front to the rear of the house are usually 2 to 3 feet wide.

Thickness is even simpler. Concrete for walks is normally $3^5/_8$ inches thick and is always referred to as 4-inch concrete. The choice of thickness is not accidental. It corresponds exactly to the height of a 2x4 lying on edge.

EXCAVATING

Start by stretching a pair of lines parallel to each other and spaced just 1 foot farther apart than the proposed width of your walk. The lines demarcate the shallow trench that will accommodate the walk.

When excavating this trench, cut through the sod and do most of the digging with a long-handled shovel. Use a square-edged shovel to make the bottom of the trench flat. The depth to which you dig should reflect the nature of the soil and your local weather, not your willingness to dig. Where the soil is dry and firm and the winters are mild, you need dig no more than 2 inches below grade, which should bring you below the roots of the grass and the topsoil and into the firmer subsoil. Where the soil is soft and wet and the winters are severe, it is wise to go down 6 inches to minimize the chance of frost heaving and cracking your work.

Digging 2 inches into the earth will let you place a 4-inch slab directly on the bare earth and finish up with the slab's surface 2 inches above grade. This is what almost everyone does. When you dig 6 inches into the ground you have to lay down some sort of a granular base 4 inches thick to bring the top of a 4-inch slab 2 inches above grade. The base can consist of gravel, cinders, crushed stone, or a combination of all these. The purpose of the base is to provide drainage and prevent water, which may later freeze and push upwards, from collecting beneath the walk.

To hold the bottom of the trench close to any desired depth, do not complete the digging until you have constructed and installed the form. Then you can place a board across the form and measure easily down from it to check excavation depth as you go. Or you can make a strike board and drag that along over the form as a guide. Strike boards are discussed in Chapter 1.

There is of course nothing wrong with going down a little deeper except that you thereby waste concrete or gravel. Neither is there anything wrong with using a little less or slightly more gravel or making the slab thicker. You should not, however,

First step in constructing a walk is to excavate a trench in which the form will be placed. The bottom of the trench should be leveled with a square-edged shovel.

make the slab less than 4 inches thick. If you accidentally dig a hole in the bottom of the trench, fill it up with gravel. Do not fill it with earth unless you dampen the earth and pound it firm with a tamper. If you do not have a metal tamper handy, use the end of a 2x4 to which two additional pieces of 2x4 have been nailed.

So much for the depth of the trench you are digging. Now for its width. This should be about 1 foot wider than the desired width of your walk. This provides space within the trench for positioning the form and driving the stakes, and permits any base material used to spread beneath the form and beyond the edges of the slab for 6 inches or so. This makes for a firmer and more stable base.

Compacting. When the soil is not overly wet and when a compromise between stability and costs is desired, the earth is not excavated to the full desired depth, but is compacted. In other words, instead of digging 6 inches below the grade and adding 4 inches of base, you remove the soil to a depth of 3 inches, compact the soil below that for a further inch or so and add only 2 inches of base. On a large walk this can represent a considerable saving. In some instances, when the soil is soft, the bottom of the trench is compacted in addition to being covered with 4 or more inches of gravel or crushed stone base.

140

Compaction is useful in that it makes soil less permeable and less liable to settle. Compacted soil absorbs less water, and therefore the damage that can be done by freezing water is less.

Compaction is easy with a gasoline engine-powered compactor, which, like all mason's tools, can be rented from one of the Rent-All shops or from a building contractor's supply company. Compaction using a hand tamper is time-consuming, when it is done effectively.

One way to determine whether or not your *exposed* soil needs to be compacted is to attempt to drive the heel of your shoe into the soil when it is dry. If you can make an impression of an inch or more, your walk will probably benefit by compaction. Incidentally, you will find it easier to compact the soil if you wet it down a little first.

CONSTRUCTING AND INSTALLING THE FORM

At this point the excavating chores have been completed. The trench has been dug to the desired depth, 1 foot wider than the width of the walk, and the sod and soil have been moved out of the way.

Now a form is constructed and installed using 2x4s on edge. Butt the end of one 2x4 to another when constructing a form that is longer than a single length. Back the joint with a nailed-on piece of wood.

Start by laying a pair of the 2x4s on edge, within the trench, parallel to one another and spaced just as far apart as you want the width of the walk to be. Drive a stake near the end of each 2x4. Raise the end of one 2x4 just high enough to bring its top edge approximately 2 inches above grade and then nail it to the stake. Repeat this operation with the other end of the same 2x4 and then with both ends of the second, parallel 2x4.

This done, measure the distance between the boards to make certain they are parallel. Measure from the boards to whatever is nearby to make certain your walk is positioned exactly where you want it, and that if it has to run parallel to a wall or meet an existing walk at right angles, it will do so accurately.

Next, extend your form just as far as necessary. Then close the ends with stop boards, which are simply boards nailed across the ends of the form, their top edges flush with those of the form. Then cut the stakes flush with the top of the form.

If the walk will be level from one end to the other, then it should be pitched to the side.

141

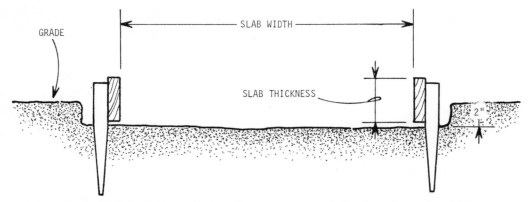

GRADE

SLAB WIDTH

SLAB THICKNESS

2"

When the walk is to be poured directly on the ground, the trench need not be more than 2 inches deep. The thickness of the slab (walk) is established by the width of the form boards. The width is established by the space between the form boards.

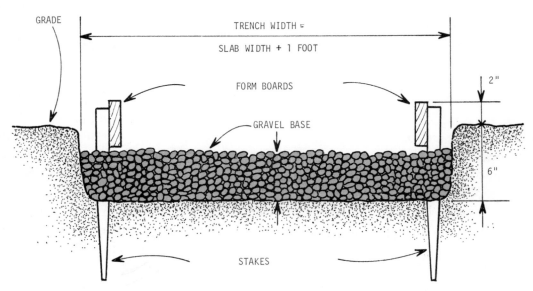

GRADE

TRENCH WIDTH =

SLAB WIDTH + 1 FOOT

FORM BOARDS

GRAVEL BASE

2"

6"

STAKES

When a granular base is to be used, the trench must be excavated accordingly. Here the granular base is 4 inches thick; trench depth is therefore 6 inches. Note that the trench is 1 foot wider than the width of the desired slab in order to accommodate stakes and permit the granular base to spread beyond the form boards.

142

After installing the form in the trench, check spacing height above grade, level or pitch.

Check the depth of the trench by measuring from the bottom of a straight-edged board placed across the form.

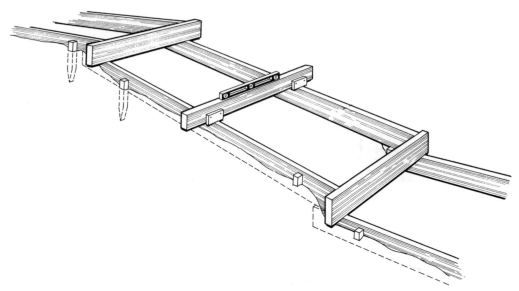

One way to construct a walk on a slope. By making the walk in a series of large steps, you reduce the pitch of the walk itself.

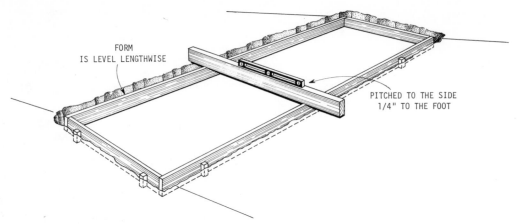

FORM
IS LEVEL LENGTHWISE

PITCHED TO THE SIDE
1/4" TO THE FOOT

If the form is not pitched lengthwise, it should be pitched to the side to drain rainwater. A pitch of about ¼ inch to the foot is sufficient.

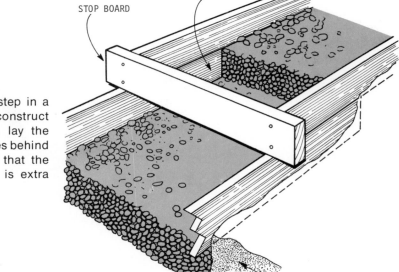

STOP BOARD

FORM

SOIL

To introduce a shallow step in a walk on a granular base construct a form as shown. Then lay the granular base 6 to 8 inches behind the front of the step so that the concrete, when poured, is extra thick at this point.

simply raise or lower one side of the form as necessary to produce a side pitch of about a quarter of an inch in either direction. This is necessary to prevent water from remaining on the concrete after a rain. Check the pitch with a spirit level placed atop a board laid across your form.

144

LAYING DOWN A GRANULAR BASE

Any granular material you may spread over the bottom of the trench, be it gravel, crushed stones, or cinders, is called a base. This is done as soon as you have completed your form. There is no special trick to laying the base beyond using small stones—say ³/₄ to 1¹/₂ inches thick—since they are easier to work with, and taking care to spread the base material evenly over the entire bottom of the trench. This means the base material will go beneath your forms all the way to the walls of the shallow excavation.

Some masons tamp the base material, but even if you use a power tamper, there is little to be gained by this. Just make the material level and be careful to leave room enough for the full thickness of concrete that you want.

You are now ready to mix and pour your concrete, but it is advisable to read on a bit. Some of the suggestions and instructions following just may be applicable to your job. If you pour now and read later it may be too late.

Use a shovel or a rake to spread the gravel or crushed stone evenly along the bottom of the trench. The granular base should be pushed under and beyond the form boards.

Save on concrete by removing all hollows in the granular base with a strike board. The board also levels the gravel to proper height.

PRECAUTIONARY MEASURES

Expansion joints. No concrete walk should be permitted to extend for more than about 25 feet as a single slab. It must be divided into two or more sections separated by an expansion joint. Otherwise, the heat of the sun will probably cause the long slab to buckle and crack along its middle.

An expansion joint consists of a resilient, bituminous fiber strip made expressly for this purpose. The easy way to install such a joint is to place a stop, which is sim-

HOW TO INSTALL EXPANSION STRIPS

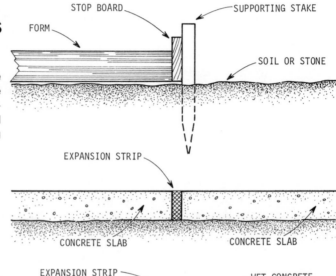

Top: Place a stop board across the form; pour the slab. Bottom: Remove the stop board, place a length of expansion strip against the hardened concrete and then pour the following section of walk.

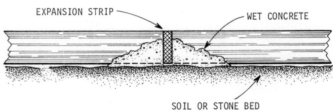

Place a strip of expansion material across the form. Hold it in position with gobs of wet concrete. Carefully pour concrete on either side.

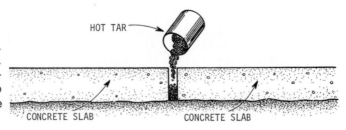

Place a stop board across the form. Pour on both sides of the stop board. When the concrete has set sufficiently, carefully remove the stop board. Then pour hot tar into the space left by the removed board.

ply a board held in place by stakes, across the form at the point where you want to install the strip. The concrete is poured and permitted to set up awhile. When it has stiffened sufficiently, the stop is removed and replaced by the strip. More concrete is then poured on the other side of the strip, locking it in place. Expansion joint strips are easily cut to length with a sharp knife. Their usual width is 4 inches, which makes them exactly right for use with 4-inch concrete.

An alternative to using the strip material is to install a stop using some globs of wet concrete to hold it in place. Then the concrete is poured on both sides of the stop. In an hour or two, when the concrete is fairly solid, the stop is pulled gently out. Later, the now empty slot is filled with melted tar.

Isolation joints. An isolation joint is merely a strip of expansion joint material placed between your walk and the side of a building or a stone wall or an existing walk or driveway. This is always done because your new walk will never move in the same direction nor at the same rate as the old buildings, walls, and driveways it may touch. Unless you have that flexible separation, cracks will develop between the new and the old structures. Water will seep in and cause problems.

To position a strip between a new walk and an existing drive, for example, simply cut the strip to size and hold it in place against the old structure with a couple of globs of wet concrete; then pour. Obviously, the top of the strip should be flush with your form.

Reinforcement. When the earth is particularly soft and wet, it is advisable to use welded steel-wire mesh to reduce the possibility of the concrete cracking, but more important, to hold the pieces together, should it crack.

6 x 6 - 10/10 is the welded wire mesh or fabric most often used for reinforcing concrete walks and driveways. It comes in a choice of roll widths, generally either 5 or 6 feet across. The size designation indicates that each wire box is 6 inches square and the wire is 10 gauge. Before you even start to think about touching a roll of welded wire mesh, put on a pair of heavy work gloves. Most of the wire is harmless, but every now and again you will encounter a razor-sharp edge, so work with care.

Use a small pair of bolt cutters for easy cutting, or a hacksaw or a large pair of electrical pliers if you have nothing else handy.

Cut the wire mesh so that when it is placed inside the form there is a 2-inch space between the wire edges and the form. If a single piece of mesh is too small to fit the space, use a second piece and join them by simply letting one piece overlap the other by a box or more. To straighten the mesh, step on it.

WHERE TO INSTALL
EXPANSION STRIPS

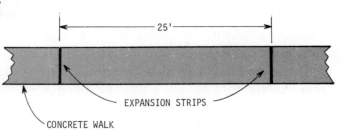

Every 25 feet (or less) of continuous
concrete walk or driveway.

EXPANSION STRIPS

CONCRETE WALK

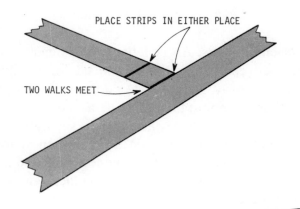

PLACE STRIPS IN EITHER PLACE

Where two walks meet. Place the
strip either at the junction point or a
short distance away.

TWO WALKS MEET

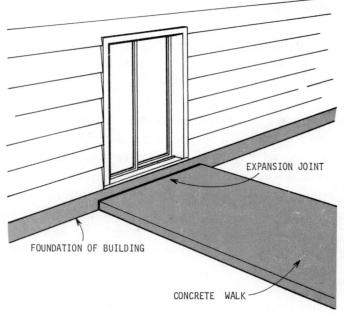

Between the end of a concrete
walk or drive and the foundation
of a building.

EXPANSION JOINT

FOUNDATION OF BUILDING

CONCRETE WALK

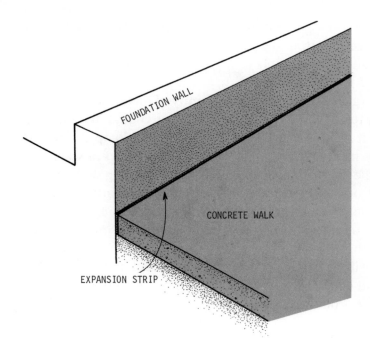

FOUNDATION WALL

CONCRETE WALK

EXPANSION STRIP

Between the side of a walk or drive and the base or foundation of a masonry wall (dry walls excepted).

Place the mesh on the base or directly on the soil if you are not using a base. Raise the mesh about 2 inches by placing a number of stones underneath. In this way the mesh will be in the center of the slab, where it is most effective for most load conditions. You can also center the mesh by pouring the concrete on top of it and then lifting it up 2 inches with the aid of a shovel or a metal hook. This method, however, is much more troublesome.

POURING AND SCREEDING

Walks are easy to pour and screed if you have three people on the job; difficult if you have only two and they are inexperienced; very difficult if there is only yourself. However, the difficulty can be greatly reduced if you break the job into small sections, say 10 feet of walk at a time. This is easily done by installing a stop board across the walk every 10 feet. If you do this, pour alternate sections, you will find it still easier.

Before you start, get yourself and your crew ready — every man, woman, and child. Each should have a shovel, an iron rake, and a pair of overshoes (the kind without buckles are best) in addition to the tools you will need for mixing, screeding, and tamping.

Start by measuring your form and estimating the quantity of mixed concrete you will need as suggested in Chapter 7. Use a $1 : 2\frac{1}{4} : 3$ mix wet enough to be easily workable, but not so wet that it slumps into a pancake when you dump it into the form. If you are working alone, mix just enough to lay down 2 or 3 feet of walk at a time. As long as you don't let the first batch set up, you can keep adding without any mark showing between pours or cracks developing there. When working with a crew, don't mix so much that you have to keep the mixed concrete waiting more than thirty minutes or so before you pour it. If you are working with ready-mix, don't let the driver pour the mix faster than you can handle it.

Hose the base down lightly if the day is hot and dry. If it is cool, this is unnecessary. Start your pour at one end of the form, and fill the end completely before you move on. Poke into the concrete around the corners with a rake to make certain the form is filled, with the concrete settled all the way down.

When 3 or 4 feet of form have been filled, screed the in-place concrete, starting at the filled end of the form and working toward the unfilled end. Screeding consists of placing a straight-edged board on edge across the form and sawing back and forth

Installing a stop board within a form. The board can be used in installing an expansion strip, or to terminate a day's work. Stakes hold the board in place.

The concrete on one side of the stop board has hardened. Now the stakes can be pulled out, the board removed, and the other side poured.

Woven wire mesh is used on soft soil to reinforce a concrete slab. The mesh can be lifted 2 inches or so off the ground with the aid of small stones. The mesh must be kept clear of the sides and the ends of the form. To join two pieces of mesh, simply overlap them; the concrete will hold them together.

A narrow form, 3 feet wide in this case, can be screeded by one person, though it is a lot easier if there are two people on the screed.

with the board while pushing it forward against the concrete. As you can readily imagine the work will proceed most easily with two people on the screed and a third on the shovel.

If the screed passes over a hollow spot, which you will recognize easily enough by the absence of screed lines, toss a shovelful of concrete into the hollow. Lift the screed and bring it back toward the end of the walk from which you started, then screed forward over the missed area again. When you have almost screeded the poured section of the walk level, pour more concrete into the form and continue screeding. Unless the form is only 5 or 10 feet long, don't fill the entire form with concrete before you begin to screed. If you do you may find the concrete at the far end too stiff to screed.

FINAL PROCESS

Tamping. In order to make certain none of the stones in the concrete rise to the surface and become visible after the job is done it is best to tamp the surface of the concrete. This can be done with the flat end of an iron rake or a commercial tamper, which covers a larger area with each stroke.

Simply hold the rake or tamper in a vertical position and drive the end down into the wet concrete for a fraction of an inch—just enough to drive the top stones down below the surface. The only requirement is that every inch of the concrete surface must be tamped.

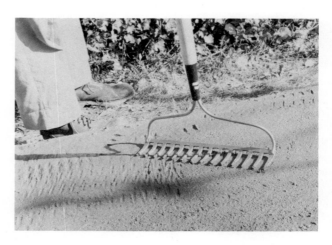

Using the end of a rake to tamp freshly screeded concrete.

An edging tool is used to round and finish the edge of a still-wet concrete slab. Work the tool backwards and forwards. Fill the low spots with a little wet concrete from which the stones have been removed and go over the spot again with the edging tool.

Edging. As soon as you have poured, screeded, and tamped about 15 feet of walk, it is time to edge that portion of the walk. This is a simple, rapid process accomplished with an edging tool. The bit, which is the underside projection, is forced between the concrete and the side of the form. Then the tool is slid along the inside edge of the form. This rounds the edge of the concrete and leaves a flat band about 3 inches wide. The operation will have to· be repeated at least once again later on.

Grooving. As soon as you finish edging, you must groove that portion of the walk that has been edged. Grooves, the deep lines you see crossing well-made concrete walks, are there not primarily for appearance sake but to induce cracks. In this way if the concrete cracks, it does so in neat square or rectangular sections, and the cracks are hardly noticeable.

Grooving a concrete sidewalk with an edging tool. The tool has been drawn across the slab in one direction; now it is being drawn across the slab in the other direction.

153

To form the grooves place a straight-edged board across your walk. Use a steel square to make certain the board is at right angles to the side of the walk. Then slide a grooving tool across the slab using the board as a guide. If you have no grooving tool, run an edger across the slab in both directions and it will form the same neat groove with two flat side bands to either side. Space your grooves an even distance apart, but no farther than one and a half times the width of the walk. If you don't groove, your walk may crack in every direction instead of along the lines produced by the groover.

Finishing. At this point you have poured, screeded, tamped, edged, and grooved the entire slab. You may be emotionally and even physically finished, but your slab isn't. It has to be given a series of treatments called finishing, and since the steps are identical for all concrete flatwork, which includes walks, driveways, patios, floors, and the like, finishing in its entirety is covered in Chapter 14. Turn quickly to this chapter because you must proceed with the finishing while the concrete is still workable.

12

CONCRETE DRIVEWAYS

All the pleasures and satisfactions you may derive from constructing your own concrete walk are to be enjoyed to an even greater extent from the construction of your own driveway. The project is larger and, to this writer, more rewarding for that reason. You may also find, that there are even greater proportional savings to be gained by making your own drive than from making your own concrete walk. Since the very size of the job is a bit forbidding, contractors feel justified in asking a little more proportionally for driveways than for walks. So when you build your own drive, your savings over local contractors' bids should be considerable.

The steps involved in constructing a driveway are very similar to those necessary to construct a concrete walk. A large, shallow form is assembled. It is filled with concrete and screeded level just like a concrete walk. There are, however, a number of important differences, which *must be reviewed* before you start or you may find yourself in difficulty.

SLOPE LIMITATIONS

Whereas most municipalities are quite unconcerned about the steepness of any walk you may choose to construct, almost all of them have strict regulations governing the slope of driveways. If your drive does not conform to their building code, you

With a little planning, a concrete driveway and parking area can be made into an attractive adjunct to a house. A travertine texture embellishes the drive and walk above, while the driveway at right combines exposed aggregate concrete with brick feature strips. *Courtesy Portland Cement Assoc.*

will be punished with a daily fine until you remove the drive or make it conform. The building department's concern with driveways is personally motivated. When a defective building and driveway are erected within the authority of a municipality, the purchaser of the building always sues the building department as the party responsible for authorizing its construction.

The code usually defines the maximum permissible slope as a percentage and sometimes also includes the driveway's profile as it crosses the sidewalk or the spot where the sidewalk may eventually be installed. Do not imitate your neighbor's drive, which may be dated, but go to the building department and ascertain all the rules and regulations. Most important, have the department approve the drive after you have excavated but *before* you pour.

Measuring slope. Slope is simply a road's drop per foot of length. A 10 percent slope means the road drops 10 feet in every 100 or 1 foot in 10. Most municipalities limit slope to 10 or 15 percent, and 15 percent is a pretty stiff grade.

To measure slope drive a long stake vertically into the ground. Measure from the bottom of the stake up the hill for a distance of 10 feet. Mark that point with a small stake and stretch a horizontal line to the large stake. Hang a line level on the string to make certain the string is actually horizontal. (Your eye is not to be trusted here.) Now measure from where the string touches the stake to where the stake enters the ground. That is your slope per 10 feet (or close enough for practical purposes). If the distance is 1 foot, for example, your slope is 10 percent. If the distance is 1½ feet, the slope is 15 percent.

Profile. The term means a side view. In our case, the profile of the driveway at the sidewalk means the shape of the drive as it crosses the sidewalk, or the place where the sidewalk may be eventually installed. Some codes insist your drive be flush as it crosses the walk. In others it can be a little lower. Some codes specify the

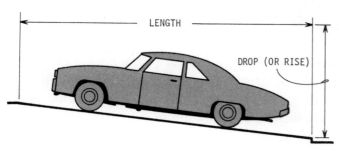

LENGTH

DROP (OR RISE)

Maximum driveway slope is isually limited by local building law. Slope is the relation between the length of a driveway and its drop (or rise) and is generally expressed as a percentage. The easy way to measure slope is to set up a vertical stick and horizontal line, as explained in Chapter 5; then use the formula:

$$\text{Slope} = \frac{\text{drop}}{\text{length}} \times 100$$

157

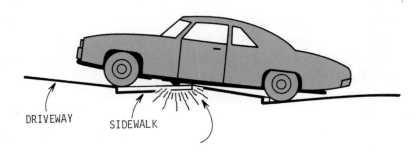

DRIVEWAY   SIDEWALK

SIDEWALK

BUMPER HITS

Why it is important to know the profile of your driveway near the street before you construct it. Top: Short cars tend to become hung up in the middle when the profile is too steep. Bottom: Long cars tend to bump their rear ends. To learn whether or not your car will be troubled, draw profile of driveway to scale and draw your car, also to scale, on the driveway.

slope of the drive where it dips to meet the road. When there is an existing curb, some municipalities insist you get permission to cut it down for the passage of your cars. They will also specify how far down you can cut the curb.

The curb is best cut with an air hammer. Hire a man with a machine to do the job for you, or rent an air hammer and do it yourself. But don't rent an electrically driven hammer, since this is too light to do the job quickly and efficiently.

Slope correction. Should you find the existing slope to be greater than the legal minimum, several courses of action are open to you. One is to lower the garage. This is not as impracticable as it may sound if you have not yet begun garage construction. In some instances it can mean the difference between being able to build on a city lot and being debarred from doing so. Another possibility is to run the drive down the hillside at an angle. This lengthens the drive without increasing the rise and so reduces the gradient. You can also secure a variance. This is legal permission from the city to construct something at variance with the code. Your chance of securing a variance is much greater if you seek it before you build rather than afterwards.

DESIGN

There are two kinds of driveways in use today. One is the single, wide-slab, and the other is the dual-runway drive.

A single-car, single-slab drive should be at least 10 feet wide for easy driving. A two-car drive should be at least 20 feet wide, with a few extra feet of width that are well worth the cost and effort. When a single-width drive widens to meet a two-car garage entrance, it is best to start widening the drive at least 16 feet away from the entrance.

When your driveway includes a curve, the radius of the circle forming the smaller, inside arc should be at least equal to the length of the longest auto that will use that driveway. For example, to accommodate a 22-foot-long car, the radius of the inner curve should be 22 feet or more for comfortable driving.

To outline a large curve on the ground, drive a peg into the ground. Tie a string to the peg. Take hold of the string at a point equal to the radius of the arc you want to form. Then holding the string taut, walk the string around the peg, using small pegs driven into the ground to mark the desired arc.

Dual-runway driveways are not very popular these days, which is a pity for they require far less material and labor than do the standard drives. Among the objections appear to be the difficulty many learner drivers find in keeping their vehicles on the tracks when backing up, and the need to remove the snow and cut the grass between the strips. But probably the main reason for their disappearance is the use of larger crews and power equipment in the commercial construction of driveways. With machines and crews it is faster to make a single slab than two slabs, even though the single slab is several times larger than both single slabs put together.

Should you opt for the good old days and construct a dual-runway, make each strip at least 3 feet wide, preferably 4. Space the strips so that their centers are as far apart as the wheels on your car; 5 feet, center to center, will be about right for most standard automobiles.

Concrete thickness. If you are reasonably certain that no trucks are going to back up your driveway, 4-inch concrete is fine. If you expect trucks, make the slab 6

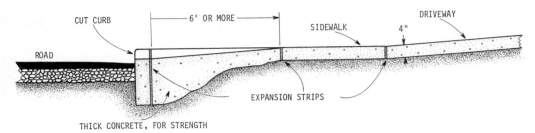

It is advisable to make concrete much thicker where it approaches the curb. This is the portion of the drive that cars and trucks use for backing up.

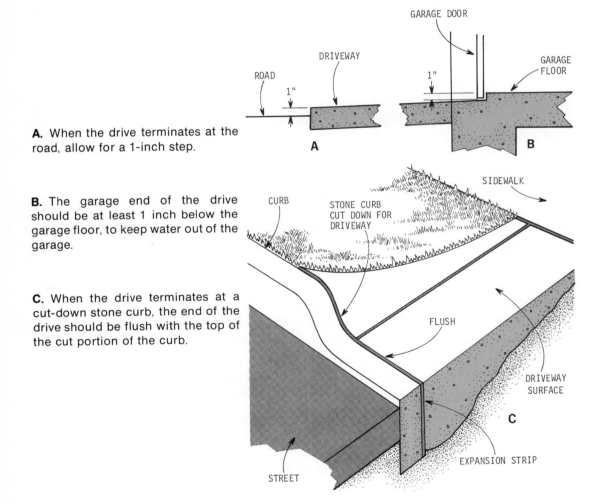

A. When the drive terminates at the road, allow for a 1-inch step.

B. The garage end of the drive should be at least 1 inch below the garage floor, to keep water out of the garage.

C. When the drive terminates at a cut-down stone curb, the end of the drive should be flush with the top of the cut portion of the curb.

160

inches thick, and if you expect no more than an occasional truck backing up a short way, make your slab a couple of inches thicker by digging down more deeply where your drive meets the road. Incidentally it is good practice to make the drive extra thick at this point because this is where the soil is generally soft, having been disturbed when the road was dug and the curbing installed.

Height. Your drive should be 1 inch below the entrance to your garage so that water does not readily flow into the garage. And bear in mind that the road end of your drive must be nearly flush with, or a little higher than, the top of the cut curb or the road. Keeping the surface of the drive 1 or 2 inches above the road helps keep out the water and dirt. The drive itself should be at least 2 inches above grade.

Pitch. Driveways, like concrete walks, must be pitched in one direction or another, to provide drainage for rainwater. In the winter a thin layer of water can quickly turn to a sheet of ice.

The need for pitch is even more imperative with drives than walks merely because the driveway is that much larger. For the same reason pitch must be part of the overall design or plan; it cannot be incorporated by merely lifting one side of the form a bit. For example, to pitch a 22-foot-wide drive $\frac{1}{4}$ inch to the foot sideways amounts to a $5\frac{1}{4}$-inch difference in height between one edge and the other. Even a pitch of $\frac{1}{8}$ inch to the foot amounts to a $2\frac{5}{8}$-inch difference.

One way to achieve this is to compromise. Use a pitch of $\frac{1}{8}$ inch and make one edge of the slab less than 2 inches above grade and the other more than 2 inches above grade. Hide the high edge by banking a little soil against it.

A better way is to construct the slab in two lengthwise sections, making the center of the slab high. Do this by building the form with a high guide board down the center and pouring and screeding one half of the form. When the concrete has hardened sufficiently to let you remove the center guide, replace it with a strip of expansion joint, and then pour and screed the second half. This method has the advan-

To pitch a wide drive, pour the slab in two sections, making a division down the middle and installing an expansion strip there. This permits the desired ¼-inch-to-the-foot pitch, which assures fast runoff and holds the slab's edges to the desirable 2 inches above grade.

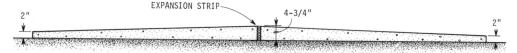

EXPANSION STRIP

4-3/4"

2"

2"

tage of dividing your large slab in two and thus reducing the chance of its cracking. In any case, your drive must be pitched sideways or lengthwise. It may be necessary to excavate to achieve this pitch.

EXCAVATING

Stretch a pair of lines the length of your projected drive, about 1 foot farther apart than the desired width. These lines will guide you when you dig a shallow trench, which, as when making a concrete walk, is the first step required in building a concrete driveway.

The same criteria suggested for deciding how deeply to dig for a walk apply when digging for a driveway. At the very least you should dig down 2 inches to get below the grass roots and possibly into the firmer subsoil, which you can easily recognize by its much lighter color. If the soil is firm, dry, and well drained and frost isn't much of a problem, you can lay your slab directly on the bare earth. If the soil is soft, damp, and poorly drained and the winters are severe, you had best dig down at least 6 inches and provide 4 inches of supporting drainage base. For a 6-inch concrete slab, dig down 8 inches.

Except for fatigue, digging by hand presents no problem. You just dig away until you reach the required level, letting the final few inches of soil removal wait until you have installed the form.

Digging by machine does present problems, in addition to paying the rental fee. You have to caution the bulldozer operator not to remove all the soil at one pass, because otherwise he will probably remove too much. This means you have either to dump extra yards of granulated base into the trench or you have to replace some of the soil and tamp it firm. Either way, it can be an expensive, time-consuming error. So it is much better that he scrape just a few inches of soil off at a time, even if the job takes longer. And, as with hand digging, delay the last few inches of soil removal until you have the form in place and can use it as a guide. This may necessitate your finishing the excavation by hand, but unless you have a really large area to level out, it is better to do it manually.

CONSTRUCTING AND INSTALLING THE FORM

Use 2x4s on edge to make a form for 4-inch-thick concrete and 2x6s for a form for 6-inch-thick concrete. Starting next to the garage, if there is one, place two of the form boards on edge, parallel to one another, with the space between them corresponding to the proposed width of the drive. Hammer stakes into the ground on the outside of the form boards and near their ends. Butt the end of one form board against the edge of the garage's concrete floor. Raise this board until its top edge is about 1 inch below the level of the garage floor.

Now, nail the form board to the nearby stake. Then, lift the other end of the form board until it is nearly level (remember, we want the driveway to pitch *down* from the garage) and nail it to a stake. Do the same with the second board on the other side of the excavation.

Next, fasten as many additional form boards as needed, end-to-end, to the form boards already in place to complete both sides of the form. At the point where the road end of the drive slopes downward to meet the curb or the road itself, use short form boards and slope them as required.

If your form is much wider than 10 feet, install a guide board down the center. use 1x4s or 2x6s on edge according to the slab thickness required. Start the guide 2 inches short of the garage end and stop it about the same distance from the road end of the drive. Use stakes to hold it in place.

Find a straight-edged board long enough to straddle both sides of the form or one side and the middle guide. Place the board across the form and a spirit level

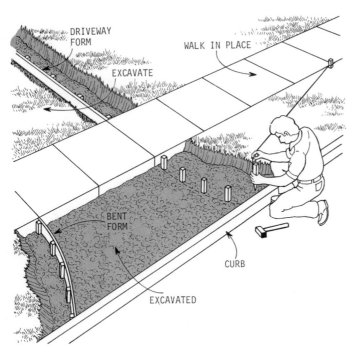

To construct a curved approach to a driveway, first excavate and prepare the drive proper for pouring; then excavate the driveway between the walk and the road. Use a long string as a compass to lay out curves for a curved form.

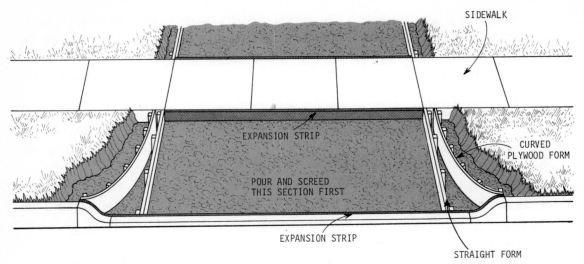

SIDEWALK

EXPANSION STRIP

CURVED
PLYWOOD FORM

POUR AND SCREED
THIS SECTION FIRST

EXPANSION STRIP

STRAIGHT FORM

When you have installed the curved form, install a straight-sided form within. Note how the ends of the straight form butt against the road and the sidewalk.

on top of the board. With the board and guide to help you, establish the desired pitch —either in one direction across the form or downward from the guide in the center to both sides of the form. If the land has a natural pitch away from the garage and the driveway is short there is no need to have a side pitch. However, if the driveway is long, it is advisable to have a side pitch to prevent a torrent from pouring down the drive during a heavy rain.

If you want the approach to the driveway, the section between the street and the sidewalk, to be curved, construct a curved form as shown in the drawings.

With the form in place and pitched as desired, use that long straight board as an easy means of checking the depth of the trench. For an even faster way, nail a second board on its side to the long straight board to make a strike board, as described in Chapter 1.

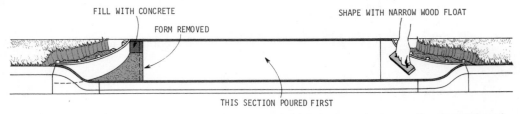

FILL WITH CONCRETE

FORM REMOVED

SHAPE WITH NARROW WOOD FLOAT

THIS SECTION POURED FIRST

After pouring the straight form, remove it and fill the space between the poured and screeded concrete and the curved form with fresh concrete. Then round and smooth the fresh concrete with a narrow float.

164

Naturally, all the high spots are removed to make the bottom of the trench reasonably level. At the same time, it is advisable to excavate down a few inches more near the ends of the drive, especially near the road, because this is where the soil may be soft from previous digging and filling and this is also where trucks may back onto the drive.

At this point you have excavated the trench, installed the form and possibly a central guide, and double-checked all your measurements to make certain the sides of the form are parallel, and that the form is where you want it and square with the garage door and the road (assuming you want it that way). If you are going to lay down a base, that is your next step.

LAYING DOWN A GRANULAR BASE

Any type of granular base material is suitable, except sand, which is too fine. Although the larger grades of stone and gravel provide a little better drainage, this doesn't justify the extra effort, so stick with stones and gravel ranging from ¾ to 1½ inches in diameter.

To estimate the quantity of base material necessary, multiply the length of the excavation in inches by its overall width in inches by *base* height in inches and then divide by 1,728 to get cubic feet. Divide again by 27 to get cubic yards because that is how gravel and crushed stone are ordered. Then add another 15 percent to make certain that you have enough.

Minimizing labor. You will discover the enormous weight of crushed stone and gravel—about 2,700 pounds per cubic yard—as soon as they are dropped off the delivery truck. Crushed stone is especially difficult to move with either a shovel or a rake. Therefore, it behooves you to plan the "drop" carefully so that hand labor is minimized.

Have the delivery truck back up the driveway between the form boards and drop the stone as it moves forward. On a wide driveway, have the truck back up twice and spread the stone in two rows. If there is a considerable slope to the drive, have the material dropped a little higher than the spot where you need it. It is easier to push it downhill than up. When the truck cannot back up the driveway, have the stones dropped off in piles as close to the form as possible. Where you have to haul the base material by wheelbarrow from the dump point to the form, make a deal with the supply yard to let the truck remain on your job long enough to dump from the truck directly into the wheelbarrow. And don't try pushing the barrows across the earth, but lay down a plank pavement on which they can be more easily rolled. Remember, you will be working with recalcitrant material that weighs 100 pounds per cubic foot. A 150-pound load is about the most anyone not used to physical work can expect to handle safely. This means each cubic yard of base will require eighteen trips. Don't dream of unloading a truck by hand yourself; get lots of help.

Spreading the base. This is simply a matter of muscle. Use a shovel, iron rake, or garden fork, whichever you find easier. Spread the stones as evenly as you can. Lay a long board across the form. Measure from the board to the base as you spread the stones to check on your progress, or use another strike board of the proper depth. Just remember, if you excavated a few inches extra at the ends of the form, don't fill the depression with base material; you want the extra-thick concrete there. And remember, too, that the base material should be spread right out to the edges of the trench, beneath and past the form boards and beneath the center guide board, if there is one.

You can tamp the gravel or crushed stone base if you want to, but as stated previously, there is little to be gained even if you use a power tamper.

The next step consists of sawing all the stakes so that they do not project above the form and guide. Then cut the expansion strip material into two lengths,

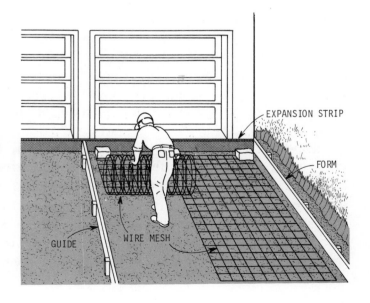

Wire mesh is rolled into position before the concrete is poured. Jump on it until it lies flat. Wire does not extend beneath an expansion strip, but is always cut a few inches short of the strip and all the sides of the form.

each a fraction of an inch narrower than the width of your driveway. Place one strip against the concrete edge of the garage floor, its top flush with the top of the form. Place a few small stones against it to hold it in position. Lay the second strip against the inside of the curb, keeping the strip's top edge flush with the top of the form. Use a few small stones to hold it in place. The two strips of expansion joint material isolate the driveway slab from the garage and the curb and reduce the possibility of the drive cracking when it expands from summer heat.

If there is neither a curb nor a garage, nail stop boards across the ends of the form. These can be 2x4s or 2x6s as necessary. Drive stakes into the earth behind the stop boards to prevent them from being pushed out of place by the pressure of the wet concrete. Incidentally, it is also advisable to install a stake to back up every 8 feet or so of form.

PRECAUTIONARY MEASURES

Expansion joints. No driveway, no matter what its width may be, should be poured as a continuous slab of concrete if it is much more than 15 feet in length. When the length is greater than this, install one or more expansion strips across the width of the form in order to divide the slab into sections, each no more than 15 feet in length.

Since the joint material is manufactured in 4-inch width, it can be used as is for 4-inch concrete. Six-inch concrete, of course, requires 6-inch-wide expansion strips because the joint must go through the full thickness of the concrete. The answer to this problem is to cut one strip lengthwise in half and place the 2-inch-wide strip, on edge, atop a 4-inch strip to make a 6-inch-wide strip.

You can hold the 4-inch strip in place with small stones or gobs of wet concrete, but you will need stakes to hold the two strips on edge, one above the other. Do not cut these stakes flush with the strips. Leave them in place until you have poured both sides of the strips, and the concrete has set up a little. Then when you pull out the stakes the top strip will remain in its proper place. Finally, fill the holes with concrete.

Reinforcement. If you want to be extra certain your drive will not crack, welded wire mesh can be set within the form at this time. The same precautions discussed in Chapter 11 involving the use of mesh to reinforce concrete walks, and the same tools and techniques, apply here. One point worth repeating is that the wire mesh should not go under the expansion joints but should be cut several inches short of each side

To be effective, wire mesh must be positioned *within* the slab. Mesh can be supported on small stones prior to pouring the concrete (above), or it can be lifted with a wire hook (right) during the pour.

of each joint. Another point, which is not a repeat, is that to save money and labor, you need only reinforce the last few yards of the drive at the road end, for it is this end that will take the brunt of the traffic. Simply lay down a short piece of mesh here and forget about the balance of the drive.

POURING AND SCREEDING

Follow the suggestions given in Chapter 7 for estimating the quantity of concrete you will need. Don't forget to add a little extra to take care of hollows in the bottom of the excavation, waste, and other losses. Use a 1 : 2¼ : 3 mix with sufficient water to make it a little more than workable—not so wet that it will find its own level, but not so dry that you can pile it up. When a shovelful slumps into a low cone about 10 inches across and 4 or 5 inches high, the mix is about right.

Decide on how you are going to handle the job and get yourself ready. You'll need the same tools you needed to make a concrete walk if you are constructing twin-drive strips. Slabs 10 feet wide and over will of course require corresponding longer screeds plus overshoes for the entire crew and a commercial tamper. You can't tamp a driveway fast enough with the back of a rake; it will set up and harden before you can possibly finish. You will find overshoes or boots an absolute necessity. Someone, and perhaps the entire crew, will sooner or later have to trudge out into the soft concrete to spread and tamp it. This cannot be done from the sides of a form 10-feet wide or more as it can from a narrower form.

If the day is hot, windy, and dry, it is advisable to wet the bare earth or the base material down with water; but don't soak it until it softens.

If the drive is fairly level, start pouring at either end. Fill the end up before you go farther. With the back of a rake tamp along the end and edges of the form a little to make certain the concrete has settled. When you have poured the full width of the drive, or to the center guide, for a distance of 3 or 4 feet, screed the poured section, then resume pouring. Don't let more than a few feet of unscreeded concrete surface remain behind you.

If the drive is steeply pitched, use a little less water in the mix and start your pour at the low end of the drive. Pour a few feet and then screed uphill. If the concrete slumps back down, wait a few minutes and then screed uphill again. Wait and repeat as many times as necessary. Do not resume pouring until the screeded section stops moving downhill. In this way you can work your way uphill, the stiffened section of the slab holding the freshly poured concrete in place.

FINAL STEPS

Tamping. A driveway is tamped in exactly the same way as a walk. The larger area to be tamped, however, makes it necessary to use a commercial tamper. The back of a rake just isn't large enough. Some masons do not tamp driveways. Their argument (beyond saving labor) is that the sight of a few stones in a driveway is not objectionable, and that the stones provide increased traction. Personally, this mason finds the sight of stones in a driveway more objectionable than in a walk, since there are more of them. And the presence of stones in the drive's surface does not increase traction but usually reduces it a little.

Grooving and edging. Driveways are usually neither grooved nor edged. However, in this mason's opinion, all driveways should be edged, since this makes for a

Finishing a large driveway with a steel trowel. Note the knee boards the workman is using to avoid marring the fresh concrete.

better appearance. Dual-strip drives should be grooved just like a sidewalk. The grooves will encourage neat cracks and help the drive look good even if it does crack a little here and there. For details on grooving see Chapter 11.

Finishing. Driveways are finished just like walks and patios and other flatwork. Finishing is covered in Chapter 14.

13

CONCRETE PATIOS

Not only will a concrete patio eliminate all the difficulties you may be experiencing with an unpaved area of your grounds, it will also provide a suitable surface for the most varied recreational activities. In addition, you will find a concrete patio much easier to keep clean than grass or a flagged surface.

A concrete patio is no more than a slab of concrete constructed exactly like a concrete walk or drive, albeit a little larger and shaped a little differently. But whereas a walk or drive merely runs between two specific points, a patio covers an area that must first be determined.

Generally a patio is placed adjoining a rear or side door to provide easy access to the interior of the building. But it can also be connected to the door by a walk or an extension of the patio slab. More important is the relation of the patio to the position of the sun at various times of the day. Will it be shaded by the building and trees in the morning when you are having breakfast? Do you prefer the shade in the morning or in the late afternoon? Do you want the patio visible to your neighbors or to the road? These major points should be settled early.

As for size, there are only a few specific guidelines. Anything smaller than a 10-foot-square patio doesn't allow for more than tables and chairs. At the same time it is a dull backyard indeed that doesn't have some grass showing, so it is wise not to cover the entire yard with a patio. And if you do leave space for grass, leave enough space to work a lawn mower, otherwise you will be clipping the grass with scissors.

A simple slab patio, built adjacent to a sliding door, provides an ideal place for summer dining and entertaining. Opening for plants can be left by installing a small form within larger form, as explained in this chapter.

Patios don't have to be rectangular. This free-form slab creates a pleasing design on the lawn.

Bear in mind that while a square or rectangle is easiest to build, the patio doesn't have to be either of these two shapes. It can be circular or free-form in outline. It can consist of a number of relatively small slabs spaced a few inches apart, the spaces filled with colored stones or grass. The patio need not be a solid slab but can have round or square openings for trees and flowers.

Height and pitch. The distance from the top of the finished patio slab to the doorway should not be more than 8 inches, which is the maximum comfortable distance one can step. If the surface of the patio will be *more* than 8 inches below the entrance, you can build the patio higher or place a large poured concrete step atop the finished patio, near the door.

The slab must be pitched away from the house, to drain rainwater and wash water. If the slab is level, water will just lie there, and if the soil subsides in the wrong direction, water will gather next to the house. A pitch of approximately 1/8 inch to the foot is sufficient.

EXCAVATING

Begin by outlining the desired patio fairly accurately with as many pegs and lines as you may require. If the patio is to have square corners, make certain the right angles are exact. If the patio is to be close to or adjoining a permanent structure, make sure the patio's sides are square with the sides of the structure. If the patio will have curved sides, lay the curves out with pegs and a line.

All this will give you a fair preview of the size and shape of the patio and its relation to the sun and the rest of the property. Try walking over the area, looking at it from different angles. Imagine using it. Make certain the patio will be where you want it and as large as you want it because removing it is more difficult than constructing it.

Move the pegs outwards by 6 inches so that you increase the overall dimensions of the outlined area by 1 foot or a little more. This entire area is now excavated to a minimum depth of 2 inches. Hopefully, this should bring you down below the roots of the grass. If the exposed soil is firm, dry, and lighter in color, indicating that it is not topsoil, and there is little frost in your area, dig no farther. If the uncovered soil proves to be dark, wet, and soft, and if your winters are rigorous, keep on digging until you are 6 inches into the earth. This, as previously discussed in the chapters on walks and drives, will enable you to place a 4-inch-thick, granular base on the soil, which when topped by 4 inches of concrete will bring the surface of the slab 2 inches above grade. Naturally, if you want your slab higher or lower, you can go right ahead. This will not adversely affect the cracking resistance of the concrete.

The bottom of the excavation should be kept as flat as possible when you are not going to use a granular base, because any change in depth makes for an equal and undesirable change in slab thickness. An overthick slab is a waste of concrete, whereas an overthin one is weak. It is less important to hold the bottom to the required depth when using a granular base because you can easily compensate for depth variations after the base material has been placed. A difference of a couple of inches in base thickness doesn't matter very much.

The best way to hold the bottom of the trench to the correct level is to let the final few inches of excavation go until you have constructed and installed the form and whatever guides may be necessary. Once these are in place you can place a board across them and measure down to the soil below. Or you can construct a strike board to use as a quick means of measuring the depth of the excavation.

Compacting. If you want to save on excavating labor and base material, and if the soil is only moderately moist and soft, dig partway down and compress the soil the rest of the way. For example, dig down 3 inches instead of 6. Compact the soil for an inch and then use sufficient base material to bring you 2 inches below grade. In this way only about half as much digging and crushed stone or other base material is necessary.

The best way to compact a large area is to rent a power compactor. The next best is to attach two strong friends to a metal tamper. As previously stated, you may sometimes find the soil soft even when you have dug down 6 inches. In such cases you have the choice of digging deeper and using additional base material, or tamping a little for added anticracking security. A little water sprinkled on the soil eases tamping.

CONSTRUCTING AND INSTALLING FORMS AND GUIDES

Obviously, some of the digging and tamping is best put off until the forms and guides are in place. They will help you to keep the bottom of the excavation level and at the correct depth. If you do dig a hollow, it must either be filled in with your base material or brought up to level with damp soil and tamped firm.

At this point you have an excavated area within an inch or two of the desired depth and about 1 foot longer and wider than the ultimate size of the patio.

When the patio is to be square or rectangular and near a permanent structure, start near that structure. As a starting point place a 2x4 or a 1x4 on its side, exactly where you want the edge of the slab to be. Drive stakes on the outside of this board, raise the board to the desired height, and nail one end to a stake. Raising the other end of the board, place a spirit level on the top edge to level the board or give it the desired pitch and nail this end to the second stake. Working from this first portion of your form construct the rest of the form.

Wood-grid patio is made by pouring concrete between wood forms, or dividers, which are then left in place. Wood should be given a coat of preservative before pouring. Here the granular base is being raked evenly between the forms.

Protect the surface of dividers that are to remain in place with masking tape. The tape is removed after the concrete has been finished and cured.

This accomplished, remeasure everything to make certain all the dimensions are correct. Check to make certain the form is square with the wall or building it abuts or is near. Check the form corners by measuring diagonally from corner to corner. If the diagonals are the same length the corners are square. Check pitch by stretching a line across the form and hanging a line level in the middle. Then lift the low end of the line while a helper watches the bubble. When the bubble is centered, the line is level. Now measure the height of the lifted end of the line above the form to ascertain the pitch for the length of the line.

In other words, if the size of the slab in question is, let us say, 16 feet, and the bubble is centered when the lifted end of the line is 2 inches above the low side of the form, the pitch is 2 inches in 16 feet or $\frac{1}{8}$ inch to the foot.

Surrounding forms. If you should want an opening in the slab—for example, to plant a tree or a flower bed—construct a form of the proper size and level it with the perimeter form. Build the form with the stakes on the inside and cut them flush with the top edge. When you pour the slab, the form will protect the area within from concrete.

Form within a form is used to provide space for a plant bed in the patio area. The form boards must be level with the boards of the larger form.

176

Guides. If the width of the slab is going to be much more than 10 feet it is advisable to install guides lengthwise within the form, thus reducing the length of the screed that would otherwise be necessary. The shorter the screed the easier it is to work with.

The guides can be constructed of 2x4s or 1x4s placed on edge, approximately parallel with the sides of the form, and with their top edges perfectly flush with those of the form. In other words, when the guides are at their correct elevation, a giant straightedge placed across the form will touch the tops of all the boards it crosses. This must be the case along the length of the form. If you do not have a straight-edged board long enough to reach across the form, stretch a mason's line to help you position the guides correctly.

Now, with the form and guides establishing the surface of the projected slab, you can do the final excavating and tamping. With a 2x4 placed on edge across the form and guides measure down to check on the progress your shovel is making, or construct a strike board by nailing a 1x6 or a 1x8 to the side of the 2x4. Nailed correctly to the 2x4, the bottom edge of the thin board will be at just the correct depth when the strike board is hung from the form and guide.

LAYING DOWN A GRANULAR BASE

When the forms are all in place and the supporting soil is at the desired elevation, it is time to place the granular base material, if you are using any. This is simply a matter of spreading the material beneath the forms and the guides all the way out to the edge of the excavation, or at least for a distance of 5 or 6 inches.

Follow the suggestions given in Chapter 7 for estimating the quantity of granular base required as well as the quantity of concrete necessary.

Use any size stone or gravel you wish, but the smaller the stones the easier they are to work with, and anything over ¾ inch will give you all the support you need. The increased support provided by larger stones isn't worth the effort.

When working on a large patio, remove one side of the form and have the delivery truck back right in and drop the material over as large an area as practical.

Spread the stones with a rake or shovel and adjust the strike board, if you have made one, to help you level and adjust the elevation of the stone base.

Some masons tamp the granular base, but as already stated, even with a power tamper, there is little to be gained in the way of compaction.

PRECAUTIONARY MEASURES

Isolation joints. Where the slab is to touch a foundation wall or any similar permanent structure, it should be isolated from this structure by an expansion joint, which as described in Chapter 11 is a strip of fibrous material made expressly for the purpose. Very simply, the strip of joint material is placed against the structure in question, its top edge level with the sides of the form. The strip itself can be held in position by small stones or a few globs of wet concrete. Later, when you pour the slab, the concrete holds the isolation joint in place. An alternative is to use an oiled 1x4, remove it shortly after the concrete has set and fill the slot with melted tar. Note that the fiber strip and the wood spacer must reach down the full thickness of the concrete. If you are pouring more than 4 inches, you have to use a wider strip or board.

Reinforcement. The considerations already discussed in regard to reinforcing walks and driveways apply to patios. When the earth is especially soft, when there is a risk of occasional flooding, when the winters are harsh and you would rather not risk the patio cracking, steel reinforcements are in order. Use welded wire mesh,

Welded wire mesh is used to reinforce patio slabs. Stones or bricks raise wire mesh to center of form before pouring the concrete.

OVERLAP

STONES LIFT WIRE
3" ABOVE BASE

MESH IS KEPT
CLEAR OF FORM

size 6 x 6 - 10/10 or heavier. Place the mesh on top of the crushed stone or gravel and clear of the form by 3 or 4 inches. Where you have to use more than one piece of mesh, have the pieces overlap by a 6 inches or more. If there are guides, remove them, leaving the stakes. Then position the wire mesh, slipping it over the stakes, and replace the guides. Or work the other way round, placing the wire mesh down and then installing the stakes and guides. Lift the wire mesh with small stones about 2 inches above the surface of the granular base. You want the mesh approximately in the center of the thickness of the slab.

Give the entire job a final review, check elevation, pitch, corners, and everything else. When you are certain that all is reasonably satisfactory, you are ready to pour.

POURING AND SCREEDING

Be sure your crew is ready. Every man, woman, and child should have a shovel, rake, and boots and for a slab of any size you should have at least one commercial tamper. You will also find a float useful at this time for doing a little hand-positioning of the concrete in the corners and along the side.

Use a moderately wet 1 : 2¼ : 3 mix. Start at the end farthest from your source of concrete supply and work with the guides, completely filling one space between a guide and the side of the form before you start on the next section. If there are two guides, it is a little easier to do the side sections first, then the middle section. Fill the end of the form, within a section, if you have them, solidly for 3 or 4 feet. Tamp the concrete down lightly with the end of the rake along the side of the form to make certain the form is completely filled. Then screed, pour some more and screed some more. Do not place the concrete down in layers. With a crew of three or four this is easy. With less hands on the job take care not to pour so much concrete that any portion of it remains unscreeded for more than twenty minutes. At the same time, if you are mixing your own concrete, don't mix so much that it waits in the pan or machine for more than thirty minutes before it is placed within the form.

Screeding, of course, consists of placing a straight board on edge across the form or form and guide and moving this screed slowly toward the end of the form while sawing it back and forth rapidly. This pushes the concrete ahead and levels it at the same time. The screed must, of course, remain tightly against the top of the form and guide at all times while doing this.

179

After pouring, start at one end of the patio and shovel or rake the concrete evenly against the wood forms before screeding and tamping.

Start screeding at one end of the form and work slowly toward the other. If the guides in the middle of the slab are to be removed, wait until after screeding and tamping when the concrete has set up slightly.

Use a trowel to fill the channel left in the slab when a guide is removed.

Use a short, straight-edged board to screed the concrete in the channel level with the balance of the slab.

To stop in the middle of the job, screed as far as you can and leave the poured edge of the slab rough. This will facilitate bonding when you resume pouring. This can be done without problem if you do not let the poured section set up. If more than an hour or two pass, wet the edge before adding more concrete to the form. Should a day or more pass, use a bonding agent on the edge before adding fresh concrete.

Concrete patios are not normally divided into sections by expansion joints, but if your slab is large and if you do not have adequate help, it may be advisable to do so. Better to have neat dividers than ugly edges left by starting and stopping the job with too great a time lapse.

FINAL STEPS

Tamping. This is most easily accomplished immediately after screeding and has to be done before the concrete sets up, at which time tamping is impossible. Where possible, reach over and tamp away: one tamp or stroke to an area is sufficient to drive the stones a fraction of an inch below the surface, which is all that is necessary.

Where this is not possible, and such will be the case for much of the slab's surface, you will have to walk on the wet concrete in order to tamp. So wait about twenty minutes and then tread gingerly on the section of concrete that has been poured and screeded first. If you do not sink more than $1/8$ inch, your judgment is sound. If you sink deeper, wait a little longer. The bottom of the pour stiffens before the top, which must be allowed time to set up.

Edging and grooving. The edges of a patio are not usually finished like those of a walk or drive. But there is no reason not to do so. It makes for a neater job. Just run an edging tool along the form as described in Chapter 11. Patios are not usually grooved, at least not for the reason of inducing cracks, as are walks and drives. Patio grooves are usually shallow and curved to simulate random flagging.

Finishing. Following screeding and tamping the surface of the slab can be finished in any number of different ways. Since finishing must be done before the concrete sets up, it is advisable to study the following chapter and prepare yourself for finishing before you pour the slab.

FREE-FORM PATIOS

Free-form is the term used to describe kidney-shaped and other randomly shaped patios. They are constructed exactly the same as the more common rectangular and square patios. The form is, of course, made of flexible material, most often plywood. When the plywood cannot be bent as desired the board is scored. A large number of saw cuts are made across the width of the board, each penetrating about half the thickness of the wood. The closer together the cuts are made and the deeper each cut, the smaller the radius to which the board can be bent.

Establishing form pitch. To level or pitch a kidney-shaped form, use a straight-edged board long enough to span the entire form. Lacking a long enough board, proceed as follows:

Excavate as suggested. Install the form and adjust its height as best you can by eye alone. Now go outside the form for a distance of a couple of feet or more. Drive four stakes into the earth to form a rectangle or a square and connect them with four lines. Now forget about the enclosed kidney, and raise or lower the four stakes until the four lines are exactly as high as you want the top of your finished patio to be, and perfectly level or pitched in the desired direction. To adjust the height of the form they enclose, stretch a line across the guide and the form, held by stakes just clear of the guide lines. Then raise or lower the form until its top edge is just below the line. Move the line from place to place until you have brought the top edge of the entire form flush with the plane established by the lines fastened to the four stakes.

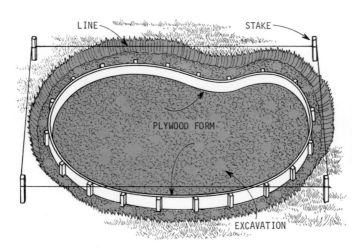

Leveling or pitching a free-form patio. First drive four stakes into the ground just outside the form. Then stretch lines between the stakes.

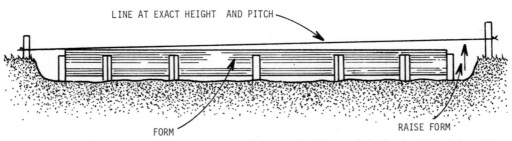

Raise or lower the lines until they are at the exact height and desired pitch of the slab you are going to pour (*continued*).

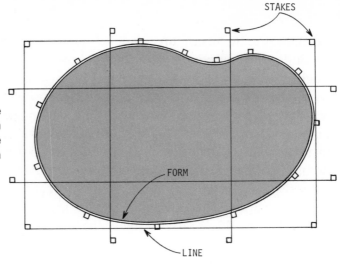

STAKES

FORM

LINE

Install four more lines touching the first four. Now raise or lower the form until it just touches the lines. The form will be at the desired elevation and pitch.

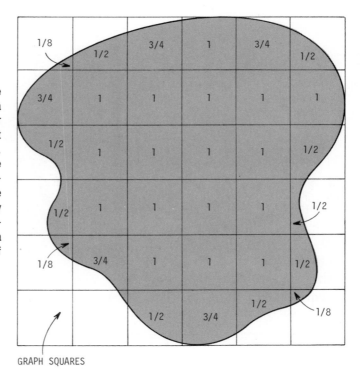

To estimate the quantity of concrete needed for free-form patios, make a drawing of the patio on graph paper as close to scale as possible. Let each box represent 1 square foot. Then estimate the portion of the boxes partially covered by the concrete. Add up all the boxes to find the total slab surface and multiply by thickness to get cubic feet. For a 4-inch-thick slab, multiply surface area by ⅓; the result is the amount of concrete you need.

GRAPH SQUARES

Estimating base and concrete. Draw an accurate proportional outline of your projected free-form patio on graph paper. Let each box of the graph paper equal 1 square foot. Count the number of boxes completely within the outline. Then count up the number of boxes that are only half within the outline. Divide by two and add the answer to the first figure. Finally tote up the boxes that are only one fourth within the outline. Divide by four and add this number to the others. The final figure roughly represents the total patio area in square feet. Since the slab is to be 4 inches thick, multiply the area in square feet by one third to get the number of cubic feet the slab occupies. Then add 20 percent for a patio under 200 square feet to get the total quantity of concrete needed, including wastage and errors. If the patio is much over 200 square feet, 15 percent additional concrete should be sufficient to take care of losses and wastage.

Follow the same procedure but multiply the area in square feet by half to get the number of cubic feet of base material necessary when the base is to be 6 inches thick. Add about 20 percent to take care of errors, wastage, and losses.

SLOPING SURFACES

So far we have considered only the construction of a patio on a level surface. We must now turn to the problems you will encounter when your land isn't perfectly level.

Start by skinning off the top layer of sod and topsoil. Check the pitch, using a line with a line level hung in its middle.

Knowing the pitch, you can decide how to correct it. Several options are open.

You can cut the high side down, making the area level, and proceed as previously suggested. You can build the low side up using crushed stone — the large grade is best here. If you do this, spread the stones 6 inches beyond the edges of the slab and let them slope gently downward. Then cover them with soil and sod. Or you can do both by removing as much of the high side as necessary and then building up the low side. This is a compromise that saves on the quantity of crushed stone you require.

When the slope is too severe for such correction, consider supporting the slab partially or wholly on piers. In other words make the slab one big porch. See Chapter 16 for data on making piers and Chapter 31 for data on making porches.

14

FINISHING CONCRETE

The final step in laying concrete is aptly called finishing. It can be accomplished in several ways depending on the surface that is desired. Float, trowel, and brush finishes are the most common.

A float finish is similar to what you would get if you slid a flat board across wet sand. The surface is smooth but has a sandy texture. This finish is used for walks, drives, and sometimes patios. It is easily cleaned, smooth, but not slippery. A trowel finish, sometimes called a steel-trowel finish, is smooth, dense, and slippery when wet. It is used for floors and patios but never for walks and drives. A brush finish is a rough, striated surface produced by drawing a brush over wet concrete. It is used on steep walks and drives to provide traction and sometimes on patios as an alternative to a float finish. A brushed appearance can also be obtained by screeding wet concrete a second time after tamping. The screeding action leaves a series of fine, low ridges across the surface. No further finishing is required.

When produced by hand, none of the finishes do very much to alter the structure of the concrete. When steel troweling is done by machine, the repeated pressure of the blades and the accompanying vibration compact the top layer of concrete, making it denser than usual, smooth, and highly resistant to wear. Machine troweling is necessary only for commercial surfaces such as those in stores and factories.

All finishing operations start after screeding and tamping, if tamping is done.

To produce a float finish it is necessary to use a float, which may be made of wood, rubber on wood, or a comparatively soft metal. If you need only to float a small surface, nail a small board on edge to the middle of a 4-by-8-inch board. Countersink the nail heads and you have a useful float.

A float finish can also be produced with a darby, which is a large, long float, and a bull float, which is an even larger version of the same tool attached to a broom-type handle.

The usual procedure is to use the bull float first to make the surface of the concrete as nearly flat as possible, then to follow with a darby and finally finish with a regular float. Where a near-perfect surface is not required, the bull float alone or the darby alone can be used.

A trowel finish is produced with a steel trowel, usually applied after the surface has been worked with a float. A rectangular trowel is used for flat surfaces and a round-edged one for curved surfaces. A machine-powered trowel is used after the surface has been floated with a bull float.

A brush finish is produced by a broom. In the trowel trade, this is called a finishing broom and comes with various types of bristles. However, you can use a street broom almost as well. For a rough, coarse finish the broom is drawn across the concrete after screeding and tamping. For a finer finish, the surface is floated before the broom is applied.

TIMING

It is very important that you start your finishing operations at just the right time. Unfortunately this cannot be specified as occurring so many hours or minutes following screeding. It varies. On a hot, dry, windy day in the sun, a stiff mix can be ready for finishing thirty minutes after pouring. On a cold day, in a cellar and poured over a sheet of plastic, a soupy mix can just lie there for hours before it is ready for troweling.

Generally, the concrete's surface is ready for finishing when the film of water and the associated shine on top of the slab have disappeared. If you become impatient and work the slab while its surface is still like mud you will bring the fines (inert powders) up, and the result will be a weak finished surface, even though you continue troweling until the slab is hard. The surface is just right for troweling when it feels like wet sand beneath your trowel.

When too much water has been used in the mix, the water that is forced to the surface by the settling stones may collect in an actual layer rather than just a film. This is called laitance and should be avoided by reducing the water quantity used in the following mix. Laitance also brings the fines to the surface and results in spalling (flaking). If this much water appears on top of your slab, brush it carefully off. Don't wait for it to dry. The concrete underneath is setting up, water or no water.

SEQUENCE OF STEPS

The slab has been screeded, tamped, edged, and grooved and is now ready to be finished. This is the usual sequence.

Floating walks. On walks and other comparatively narrow slabs of concrete, use a darby or a float. Swing the tool in circles, holding the leading edge of a new tool up a fraction to keep it from digging into the concrete. (No need to do this with old tools; their edges are rounded.) Keep moving and float the entire slab. Quit when the float no longer leaves visible marks. At this point there is nothing to be gained by continued floating.

Floating driveways and large surfaces. Begin with the bull float. If possible make your strokes intersect, working a while from the end of the slab and then from one side. When there are no hollows and no missed spots, you can quit without more ado if the appearance is satisfactory, as well it may be on a driveway. For a smoother surface, free of all bull-float marks, you have to follow up with a darby or a float. Both these can be moved in circles, producing the desired polishing action, which cannot be easily done with the bull float.

Where you cannot stand on the turf and reach over to work the concrete, you have to walk on it. Do this by laying down a path of flat, clean boards, and picking them up behind you as you work.

If the slab is small, a float can be used to finish the surface. Work the concrete with a circular motion and quit when the surface is fairly smooth.

When the surface to be finished is fairly large, use a bull float (above) or a darby (right). They are faster and produce a flatter surface because they cover a greater area. *Courtesy Portland Cement Assoc.*

When you can't reach a portion of the slab without walking on it, use boards to step on.

189

To use the float or the darby, of course, you have to get down on your knees. To avoid getting housemaid's knee, use kneepads. The masonry supply yard should have them.

Use the darby and float following the bull-float treatment exactly as described above. Keep moving the float around until its passage leaves no visible marks. Don't forget to go over the edges and grooves again.

Steel troweling. It is best first to level the surface with a float, then to swing the steel trowel in the same circular, pressing, and polishing motion already described. Keep the leading edge elevated slightly or it will dig in. Continue working until the surface is smooth and dense and you cannot see the individual grains of sand. For a slightly better surface, sprinkle with water and trowel some more. Complete the job by running over the edges and grooves again.

To make a smoother, less porous surface than is possible with a float, follow the float with a steel trowel. Work the trowel in a circular motion, keeping the leading edge up a little to prevent it from digging into the concrete. If the concrete is hard, sprinkle a little water on it.

When you have completed finishing a section of the slab, redo the edges and grooves. This is the last step before curing.

Large surfaces cannot be smoothed with a hand trowel. The concrete will set up and harden faster than you can keep up with it. The only practical method, except to pour and finish small sections at one time, is to rent a power trowel. *Courtesy Gold-blatt Tool Company*

Using a power trowel. The power trowel is usually applied after the surface has been floated and a little later than you would start working with the steel trowel by hand. The reason is that you have to stand to use the machine and there is no point in troweling when your heels dig into the concrete. The procedure is exactly the same. Just keep the machine working and continue sprinkling water for a few minutes after you have a hard surface and can see no change.

Brushing. To produce a deeply striated surface, draw a suitable brush over the concrete as soon as surface water and the associated shine disappear. Draw the brush straight across the slab, preferably in a line across normal traffic flow, in parallel strokes but without overlap. Don't push the brush; it may dig in.

To produce a striated surface, draw a brush across the concrete after it has just been floated and is still soft. Follow with an edger.

191

For a somewhat softer finish, draw the brush a second time in the same direction across the same area. This somewhat reduces the depth of the striation.

For an even less pronounced striated finish, float the surface first, then apply the brush without any pressure but its own weight. Complete the job by going over the edges and grooves.

Washing your tools. If you are not going to use any of the finishing tools mentioned a second time within less than ten minutes, wash the tools thoroughly. You can't produce a smooth surface with a rough tool. You'll be surprised at how well concrete can stick when you don't want it to.

Curing. The final step in finishing concrete is curing. On flat work, curing consists of protecting the surface from sun and wind. If the slab's surface dries before it has cured, it will be soft and weak. It may powder off and it may spall (flake). Curing can be accomplished in any of several ways.

One is to cover the otherwise finished slab with a layer of wet straw, wet newspapers, canvas or burlap, polyethylene, or even wet sand. Except when using the canvas or plastic, keep the covering materials moist by frequent but gentle wetting. Remove after two or three days.

If no covering materials are available, keep the slab's surface moist with a fine spray of water. A garden soaking hose (a hose with holes in it), hole side up, is fine after the concrete has hardened enough not to be affected by the movement of the water. Place the hose on the high side if there is a pitch. If you don't have a soaking hose, just keep spraying until sundown. Resume spraying at daybreak, if the day isn't cloudy or it doesn't rain.

A light rain will eliminate the need for artificial wetting. A heavy rain immediately after troweling can pit the surface of the slab. Protect it as best you can with canvas, plastic, or newspaper. If your work suffers a really heavy downpour before you can protect it, and if the surface is visibly pitted, try using the trowel again. If that has no effect, mix 1 part cement with 3 parts sand and a little water. Wet the slab and spread the mixture as thinly over the slab as you can. Then trowel it in place. The results won't be beautiful, but the pitting will be gone.

Should you want a better job, apply one of the bonding chemicals to the concrete's surface and then lay down a ½-inch-thick layer of the cement-sand mix. Level and repeat your finishing operations. This is the way slabs were finished in the old days: a thin cement-sand mix on top of a cement-sand-stone mix.

Concrete is cured by keeping it moist. This can be done any number of ways. The least expensive method is to cover the surface with newspapers. Keep the paper dry in the winter, when you need its insulating properties. In summer, when the need to conserve moisture is more important, wet the paper.

Plastic sheeting is also used to seal the surface and requires no water. Sheeting must be laid flat, overlapped at the joints, and anchored at the edges.

SPECIAL FINISHES

Divisions. You can change the appearance of a smooth concrete surface by dividing it into squares or diamonds with a groover. This is of course done shortly after screeding and tamping and repeated following the finishing operation.

You can make the surface divisions deeper and broader by using strips of wood, which are temporarily embedded in the wet concrete after screeding. The strips should be oiled to keep the concrete from sticking. They can be up to 1 inch thick and 2 inches wide. Thicker strips go more deeply into the concrete and are more difficult to remove without breaking the edges of the concrete, and the wider channels they create are hazardous. The strips are pushed into the concrete after screeding and are carefully removed after finishing.

An informally flagged surface can be suggested by using odd-shaped, curved pieces of wood as temporary dividers. When they are removed, they leave channels of varying depth and width, making the slab visually more interesting than if it were simply smooth.

Strips of wood, pressed into the concrete while it is still wet and then later removed, will produce surface divisions that improve the appearance of a large and otherwise unadorned slab. *Portland Cement Assoc.*

Use a piece of small-diameter pipe to simulate random flagging.

You can also imitate random flagging patterns with a groover by making short, curved, and angled grooves on the surface of the concrete, simulating joints between random flags. If you don't want the smooth side bands created by the groover, make the grooves with a piece of small-diameter pipe dragged through the soft concrete.

Where there is no frost to speak of, you can divide your concrete slab into regular or irregular shapes by permanently installing wood dividers. These can be strips placed in the form before pouring or driven into the concrete immediately after pouring. Their top edges must, of course, be flush with the concrete's surface.

A secondary advantage of breaking the larger form up into small, permanent forms is that you no longer need to finish the entire slab in one day; you can do one small form at a time as convenient.

Embossed surfaces. A few masonry supply yards and a few rent-all shops stock masonry stamping tools. Somewhat similar to rectangular cake tins, they are inverted and pressed into wet concrete to form raised shapes or bosses that may look like the top portion of bricks or Belgian blocks or other regular shapes. You can also make do with anything else that strikes your fancy as a pattern maker: pie tins, plates, coffee cans, etc.

You can turn some of the molds over and produce reverse embossed patterns of various shapes and designs. (Do not make them very deep.)

When planning to do any of this special finishing, use a 1 : 2¾ : 3 mix with ½-inch or smaller stones. This mix, being high in sand, lends itself more readily to surface forming than other mixes.

Keep your molds clean; wash them free of adhering concrete as you work. Take great care to make each impression sharp-edged and clean and to keep the patterns in line and square with the slab's sides and ends. Slight errors in this kind of work show up very readily and detract from the overall appearance of the job far more than you might imagine.

Small textural variations. Here are some suggestions for varying the normal surface texture of finished concrete to a small but highly visible degree.

When you have completed your troweling, spread a little rock salt over the surface of the finished, but not cured, concrete. Use large pieces and keep them somewhat separated. When the slab is hard, wash the salt and water solution out, and the result will be small, irregular holes in the surface.

When you have completed your work with the float, and the concrete is smooth but not hard, spread a thin layer of coarse sand evenly over the surface. Work the sand partly into the concrete with a float. This will leave a sort of sandpaper finish. Use colored sand for a more striking effect.

Use a brush to produce striated squares, with the lines in each adjoining square running at right angles to each other. Do this before or after float finishing.

With a 1- or 2-inch-wide float (a smooth board on edge) smoothen the edges of the striated squares. This will produce a soft, basket-weave effect. Use wood guides to help you make the squares even and of equal size.

You can form circular striations by swinging a brush in circles or semicircles. Again, this may be done before or after float finishing.

Use the float a little early. Swing it in a series of arcs or circles. Do not follow up with a crossing motion, but let the original broad strokes remain.

Place coarse, wet burlap on top of the concrete after it has been float finished. Using one or more planks carefully press down on the burlap, taking care you press evenly everywhere and the burlap is smooth and uncreased. Lift the burlap gently. Its texture will remain imprinted in the concrete.

Exposed aggregate finishes. There are two ways to produce this finish. One is to use an attractive aggregate in the mix in place of the usual stone. The mix is kept a little on the stiff side, screeded but not tamped. After it has set a little, the surface of the concrete is brushed and hosed down at more or less the same time. The brush loosens the top layer of concrete (cement) and the water washes it away. Thus the tops of the aggregate are exposed.

The second method consists of pouring and screeding as usual, then spreading the desired aggregate evenly over the surface of the wet concrete and tamping it in. After this the surface of the concrete is given the same brush and wash treatment to expose the tops of the aggregate.

The first method never fails to produce an attractive surface. However, unless you use bank-run gravel (round stones of varied sizes) or selected gravel (round stones of one size), the cost of making an exposed aggregate surface can be fairly high because the really attractive stones are expensive.

The second method is less costly in material, but requires more work, and you run the risk of some unwanted, unattractive stones appearing. This can be reduced by placing so many attractive stones on top that the others cannot come to the surface. But in that case the cost of the stones can again be considerable.

For the second method use the 1 : 2¾ : 3 mix because it is high in sand, along with ¾-inch stones and a little more water than usual. Pour and screed a small section of the slab. Spread your stones as evenly as you can across the freshly screeded slab. Use a shovel for small stones; place the large stones by hand. Leave no space uncovered. Press all the stones into the concrete with the aid of a float. Then lay a flat plank on top of the stones and pound on it with a small sledgehammer to secure a flat surface. Quit pounding the plank when the tops of the stones are just about level with the surface of the concrete. Then wait until the concrete has almost set. At this point the water will be gone from its surface, and it will be firm, but not hard. Now, brush the surface lightly with any convenient brush, and at the same time, hose the concrete gently to remove the material you loosen. When you have exposed sufficient aggregate — that is to say, when you are satisfied with the appearance of the slab — quit brushing and hosing. Cure the slab in the usual way.

If you want grooves and edges, do the grooving and edging before you place the aggregate, taking care to keep the stones within the borders provided by the groover and edger. Then, when you have finished pounding, go over the grooves and edges again. With a little luck and lots of care, you can retain the groove and edge bands.

Variations on a theme. The same technique can be used to implant flags in the surface of the concrete. As before, the concrete is poured and screeded, but this time it is also tamped. The flagstone that is to be implanted is laid on the slab in the

LAYING EXPOSED AGGREGATE CONCRETE

1 Concrete to be given an aggregate finish is first poured into a form, struck off, and finished with a bull float or darby. The surface should be ⅜ to ½-inch lower than the forms to accommodate the extra aggregate.

2 Spread the aggregate uniformly across the concrete surface until the entire area is completely covered with a layer of stones.

3 To embed the aggregate in the concrete, first use a bull float or a darby. Then for a final embedding, use a hand float until the aggregate is concealed beneath the surface.

4 Allow the concrete to set up until the surface can bear the weight of a man on knee boards with no indentation. Then remove excess concrete by brushing the slab lightly with a stiff, nylon-bristle broom.

5 To expose the aggregate, spray the surface with water. Here a special spray-broom is used to clean and brush the surface. Spray and brush until the water runs clean and there is no more cement film on the aggregate. *All photos from Portland Cement Assoc.*

199

desired position and its outline scratched in the concrete with the point of a knife or similar tool. The flag is removed and a shallow, form-fitting hollow is scooped out of the surface of the wet concrete. The flag is dropped inside and pressed down flush. If the hollow is too deep, add some wet concrete under the flag and then press the stone down until it is flush with the surface of the slab. Once implanted, the flags are wiped clean with a wet rag, and the rest of the surface is finished with a float or steel trowel. The resulting surface is flat and smooth but with more visual interest than that of plain concrete.

The flags can be placed at random or used to lay out a path from one point to another, or to mark the corners of a playing area, as for example a volley-ball court. Cut the flags to the proper shape, position them carefully and you can have a permanent shuffle-board court.

Another method of making exposed-aggregate concrete is simply to scatter small stones over the surface and pound them in with a 2x4 and small sledge.

A concrete slab can be given a perforated finish by sprinkling rock salt across the surface before the newly poured concrete sets up. *Courtesy Portland Cement Assoc.*

Another variation that results in a finely pebbled surface consists of covering the freshly poured concrete with a layer of screened gravel. Some yards stock gravel under ½ inch in diameter. If this is not available, and you are using bank run gravel on the job, you can screen out the small stones yourself without too much difficulty. These are tamped into the concrete and given a float finish. The surface is then gently sprayed with water. With a little care a finely pebbled surface can be obtained in this way.

Cleaning. Whenever you expose the aggregates in concrete, or embed flags in concrete, a film of cement is always left behind after some of the concrete has been washed away. This film masks the colors of the stone and dulls the appearance of the finished slab. Try removing it with a stiff brush and warm water the day after finishing. If that is not effective, try muriatic acid and water—One part acid to ten parts of water by volume. *ALWAYS* add the acid to the water. Use elbow-length rubber gloves and lots of care. Give the acid an hour or so to do its job, then flush with plenty of clean water. Or use a commercial cleaner such as Brick Bath. Cure afterwards as previously suggested.

COLORING CONCRETE

Concrete can be colored in three ways: dusting, integral coloring, and staining. A wide range of colors are available, the greens and blues costing more than the other colors. They can be used with any of the cements, but for the cleanest, brightest results, use white cement in the mix.

Dusting. This is done by spreading (dusting) a thin layer of dry concrete pigment over the surface of the slab immediately after it has been lightly floated. The color is worked into the concrete with the float and then with a steel trowel, if so desired. The slab is then cured in the usual way.

With experience and a sure hand, coloring by dusting is fast and inexpensive. Without experience, you will run into one major problem: spreading the material evenly. Several techniques have been developed to overcome this obstacle. Here are some of them.

Cover the areas not to be dusted with layers of wet newspapers or boards.

Mix the dry color with dry cement, or cement and sand. Use white sand for brighter results. Adjust the ratio of color to cement to suit your taste. Mix up a small sample and dry it in the oven, if you can't wait, to see what it will look like. Bear in mind that the less color you use in relation to the other ingredients the more evenly the color will be spread, but the lighter it will be. You can also purchase dust-on colors already mixed with cement and sand from masonry supply yards.

Apply the color in two layers.

Use an old hand sprayer that can handle powder to secure even dispersal. (Some garden sprayers are made for fine powders.)

The best technique of all is to practice awhile on some homemade poured concrete flagstones, or even on wet newspaper, to steady your hand and eye.

As soon as you have lightly floated the slab, spread the pigment as evenly as you can. You don't need much, just a covering layer. Give the pigment a few moments to absorb moisture, then go over the surface gently with a trowel. If a lot of water comes up, as it may, and turns the pigment into a mud, stop and let it dry a while.

After floating, follow up with another light dusting and some more floating. Still later, when the slab has set a little more, float some more. You can use a steel trowel afterwards if you want to.

When you cure the slab do not use a sheet of plastic or canvas. Condensation beneath a watertight covering leads to spotting.

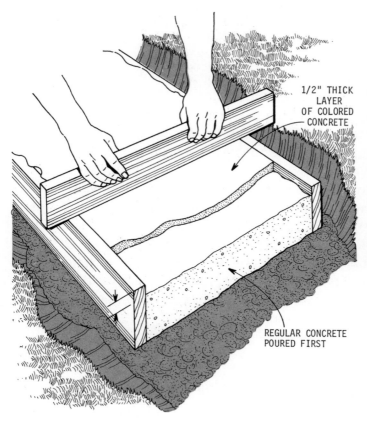

1/2" THICK
LAYER
OF COLORED
CONCRETE

REGULAR CONCRETE
POURED FIRST

The two-layer method of coloring concrete. The bottom layer is poured first and its surface is held approximately ½ inch below the desired finished surface level. The missing ½ inch is then filled with a layer of colored concrete, which consists of white cement, white sand, and the color, but no stones. The second layer is screeded and floated as usual. There is no need to tamp, as there are no stones near the surface.

Integral coloring. This method produces the most even results. The color may be added in one of two ways. One is very easy but expensive. The other is far less costly but involves more work. The easy method involves a single step, the less expensive method, two steps.

The one-step procedure. Throw dry concrete pigment into the concrete as you mix it. Just be careful to throw exactly the same quantity of pigment into each batch of concrete. Since you will need from 5 to 10 pounds of color for each bag of cement you use (5 to 10 percent of the weight of the cement used), and since you should use white cement for brighter results, the cost of colored concrete made by the one-step method is not insignificant.

Naturally, you will need less of the dark coloring than the light. Again, it is advisable to make up a sample batch and dry it to see exactly the color that will result before you commit yourself to coloring the entire slab.

The two-step procedure. Pour your form with the regular mixture of concrete, but do not screed. Instead, strike it off approximately ½ inch below the desired finished surface. This is the first step.

The second step consists of topping the poured concrete with a colored layer of "concrete," which is actually cement and sand; no stones. The topping is best laid down immediately after striking, or at least on the same day. If more time must pass, roughen the surface of the concrete slightly with a rake. Wet the concrete down, and if it isn't still green, apply a layer of any of the bonding chemicals such as Weld-Crete or Permaweld-Z.

Pour the colored mix onto the slab, screed and finish the normal way.

Top mix. Use white cement for best results. Mix it with sand in a ratio of 1 part cement to 3 parts sand by volume. Add color on the basis of the weight of the cement: 5 to 10 pounds of color for every bag of cement used.

A bag of cement plus 3 cubic feet of sand will give you about 2½ cubic feet of mix. A ½-inch-thick layer amounts to 648 cubic inches per square yard. Thus 2½ cubic feet of mix will cover 6.66 square yards of surface (½ inch thick). Figure 6 yards, to allow for waste and hollows in the slab's surface.

Staining. Concrete is easily stained with an organic stain, a solvent stain, or an oxidizing stain.

Organic stains. These stains are preferable to the solvent stains because the organic solution works by changing the concrete chemically, thereby producing a more permanent color than the solvent stains. However, organic stains should not be used on patched concrete, as the stain affects concrete of different ages and different composition differently, resulting in noticeable variations in shades. Organic stains produce brown, green, black, rust, and beige colors. The stain is simply brushed on the finished concrete surface. See the container for specific manufacturer's instructions.

Solvent stains. These stains merely soak into the pores of the concrete. Essentially, solvent stains are very similar to wood stains. However, it is best to use the stains formulated specifically for use with concrete. Though the stain is usually just brushed over dry, finished concrete, it is wise to read the label. The manufacturer may have some special instructions.

Oxidizing stains. These are old-time solutions that you have to mix yourself, and which you should try on some sample pieces of concrete before you cover your entire walk or patio.

To produce buff and reddish brown, mix ferrous sulfate (sulfate of iron) with water at the rate of two full tablespoons of chemical to one pint of water (two standard glasses). Stir until the chemical is completely dissolved. Then with a stiff brush spread it as evenly as you can over freshly finished concrete. For best results the concrete should be just a little too hard to be marked by the brush.

Take care when applying the mixture. It will permanently stain your clothing and kill whatever grass it touches.

The concrete will turn green at first, then brown. The intensity of the color will depend on the strength of the solution and thickness of application, so be careful not to overlap your brush strokes or some areas will be darker than others. If you wish, you can go over the entire surface a second time to darken the color. This solution can be used on concrete made with regular gray cement.

To produce green, mix copper sulfate with water in the ratio previously mentioned. Apply the same way and with the same care as soon as the concrete has been finished and has hardened sufficiently not to be marked by the brush.

Cure the stained concrete the usual way.

FLAGGING CONCRETE

There are many reasons for flagging concrete. None of them are utilitarian, meaning flagging a slab of concrete does little to extend its life or usefulness. But beauty has its own rewards, and to this mason, at least, nothing is uglier than an expanse of bare concrete slab.

This chapter, therefore, will tell you how to permanently cover any concrete slab—a patio, walk, or driveway—with flagstones. The flags can be any kind of stones you wish, including quarry tiles, which are not stones at all but pieces of fired clay.

To flag concrete the surface of the slab is covered with a layer of mortar, then the flags are pressed lightly into the mortar, one after another, each flag in its planned position with a small space between them. To make the layer of mortar smooth and level it is screeded with the aid of guides much like a cellar floor is laid down and screeded. If some flags are thicker, they are pressed down more deeply; if thinner, raised with a little added mortar. After the mortar has hardened, the spaces between the flags are filled with more mortar, making the flagging a single, solid, continuous waterproof layer up to several inches thick.

This attractive patio was made by flagging a concrete slab with precast concrete paving blocks. The blocks are available in many shapes, colors, and textures.

FLAGGING MATERIALS

With the exception of quarry tile, all the flagging materials commonly used are covered in Chapter 1.

Quarry tiles are made of fired clay just like bricks. The most common sizes are 6 by 6 inches and 9 by 9 inches, ⅝ inch thick. Because of their comparatively small size, they are the most difficult of all the paving materials to lay down smoothly and evenly. Use the largest tiles you can find, hold the joint width to a minimum and your job will be simplified. But remember, to be beautiful, quarry tiles must be laid perfectly smooth and perfectly even.

SELECTING AND FITTING A PATTERN

Assuming you have decided on your flagging, the next step is to select a pattern. Bear in mind that the simpler the pattern the easier it is to install and the better it generally looks.

To save time and reduce the cutting, lay your stones out on a flat surface exactly as you want them to appear when the job is done. Take care to space them accurately. Do not depend on the yard supplying you with exact-size stones unless you have them cut to size. There will be slight size variations that may end up as large variations before the entire area is flagged. So lay out the flags and measure them up first. Then number them so you can easily retain the same order. Now you know exactly how large your slab needs to be. If the slab is in place, you have to work backwards. Measure the slab and fit standard size stones in place by varying the spacing between them.

You can have the supply yard cut the flags to your sketch. In such cases be certain to supply them with the spacing you want and specify the method by which the flags are to be cut. The cheapest is snap cut, which leaves a rough edge. For a smooth edge the flags have to be sawn. This by far is the more expensive method. Of course, you can always cut the stones yourself, as previously discussed.

PREPARING THE SLAB

If you are going to construct a new slab, proceed exactly as discussed in previous chapters. Make the slab's size conform to your pattern requirements. Do not forget to give it a pitch. And bear in mind that the flagged slab will be thicker than the base slab by 1½ inches plus the thickness of the flags, so dig accordingly.

Pour, tamp, screed, and cure the concrete. Do not finish beyond screeding. You want the rough surface left by screeding.

If you are going to flag existing concrete and the surface is clean, just wet it down before proceeding and apply a coat of Weld-Crete or any other concrete-bonding agent to the area you are going to work on immediately. If the surface is covered with paint or anything similar, all of it must be removed. The best way is to use an air hammer and just chip a fraction of an inch off the surface. Paint remover followed by a solvent is not nearly as good because a film always remains and interferes with bonding, which means that the top layer of flagging may lift up when hit by frost.

You are now ready to flag the concrete. There are a number of methods or techniques that can be used to do this. The following method is a bit tedious but has the virtue of being almost infallible. When you gain experience you can eliminate some of the steps.

CONSTRUCTING THE FORM

Position 1x6 or 1x8 boards (yellow pine roofers are good enough) on edge, tightly against the sides of the concrete slab, thereby enclosing the slab in a wooden form. The top edges of the form boards should be flush with the surface of the flags to be laid. In this example the flags are 1 inch thick and will be laid atop mortar $1^5/_8$ inches thick. Therefore the form boards should project $2^5/_8$ inches higher than the top of the slab. With a spirit level, make certain the form boards are level or are following the pitch of the slab.

Next lay out the flag positions by marking the form boards. Bear in mind there will be spaces between the flags, so allow for the spaces.

Later, when you are actually laying the flags, drive nails into the marks and stretch lines across the form in two directions to check on your progress. In this way you will quickly see whether or not you are holding closely to the desired pattern. If you aren't the trouble may be due to changes in the size of the flags or changes in the spacing between them.

SPREADING THE MORTAR

To make certain the mortar on which the flags will be laid is smooth, level and exactly the right thickness—in this case $1^5/_8$ inches—2x4s are laid on their broad sides on top of the slab and are used as guides for screeding.

The first 2x4 or guide is positioned a few inches clear of one end of the form. A second 2x4 guide is laid down roughly parallel to the first and spaced 2 to 3 feet away. This may be followed by a third and fourth guide similarly positioned. It isn't necessary to use more guides at one time. Each 2x4 does not have to reach from one side of the form to the other. You can use small pieces and let their ends lap. But the pieces of wood must be perfectly flat.

Mix up a small batch of mortar consisting of 1 part cement and 5 parts sand. Mix the two ingredients together thoroughly and then add a little water at a time, testing as you go. When you can form a ball of mortar without it dripping or falling apart, you have just the right quantity of water.

The mortar ready, wet the top of the slab where you have placed the screeds. If the slab is old or has been troweled smooth, use a bonding agent to wet the surface.

Next spread the mortar over a 3- to 4-foot square area adjoining a corner where you have placed the guides. Screed the mortar level with the tops of the guides by sliding a straight-edged board across the tops of the guides. Fill in any hollow area. Then remove the guide lying next to the form and fill that space with more mortar. Now, with a delicate touch, screed the newly placed mortar level with the other guide, making certain there is no space between the mortar and the form boards adjoining the slab.

FLAGGING A CONCRETE SLAB

1 Enclose the slab within a form of boards held in place by stakes.

2 Adjust the height of the form to equal the thickness of the guides. The top edge of the form will thus be flush with the tops of the flags.

3 Level the form or, if it is necessary to pitch the slab, check the incline.

4 Lay out the flag positions on the sides of the form. Use these marks to place flags in proper positions.

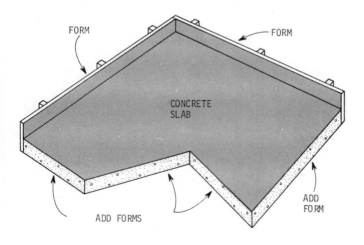

FORM

FORM

CONCRETE SLAB

ADD FORMS

ADD FORM

5 If the slab is irregular, construct the form in sections, installing two or more forms at a time, as needed.

6 Place one 2x4 guide a few inches away from the form and a second guide parallel to the first about 2 to 3 feet away. After wetting the surface of the slab, spread the mortar over the wetted area.

7 Screed the surface of the mortar flush with the guides.

8 Remove the guide nearest the form and fill the trench with mortar.

9 Using the surface of the smooth, in-place mortar as a guide, gently screed the mortar you have just added.

10 Wet the surface of the mortar and the underside of the flag. Lay the flag in place, using the edges of the form as a guide, but do not allow it to touch the form.

11 After you have placed the second flag in position, lay a 1x2 spacer on top of the guide. Then place a straightedge across the flags, spacer, and form. If the straightedge just clears the flags, they are properly positioned.

213

12 Wet the underside of each flag and the surface of the mortar.

13 If a flag is too high, tamp it down or remove some of the mortar from beneath it.

14 When you have flagged the mortar in one section, spread mortar in the next screed it. Remove the nearest guide and after filling in the trench left behind, screed the surface.

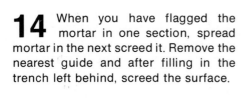

15 Every now and then, check your work by laying a long straight-edge across it.

16 Allow several days to pass before you tackle the joints. Then fill them with the same mortar. Force the loose mortar tightly into the joint to make a smooth, finished surface.

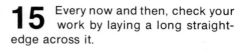

LAYING THE FLAGS

Take a masonry brush (a paintbrush will do) and sprinkle a little water on the underside of the first flag and on the mortar on which the flag will rest. Lay the first flag in place using the corner of the form as a guide. Do not let the first or following flags touch the form. They will push it out of shape. Tamp the flag gently into the mortar with the aid of a short 2x4 and a small sledgehammer. A few taps is all that is necessary. Following the same procedure, lay the second flag in place.

Next, place a 1x2 on its flat side on the guide near the two flags. See that the top of the guide is clean. The 1x2 serves as a spacer. Its top and the top edge of the form will be exactly level, assuming that the form is at the correct height.

Lay a straight-edged board across the edge of the form and the spacer above the flags. The straight edge should just clear the tops of the flags. If one or both of the flags touch the straightedge, place a short piece of 2x4 on the flag and pound the flag gently into the mortar. If gentle pounding doesn't do it, lift the flag and, with a trowel, scoop up some of the mortar. Smoothen the mortar, drop the flag and try the straight-edge again. On the other hand, if the flag is too low, remove it, add some mortar and try again.

When you have laid two flags satisfactorily, use a stick of the proper width to help you space the next two flags. When you have covered the spread mortar with flags, mix and spread some more mortar. When you have flagged one end of the slab, remove and replace the guides on the adjoining portion of the slab. Spread some more mortar and continue to lay the flags. Every once in a while, stretch lines across the flagged patio to check on the pattern. In this way you will be able to hold the flags properly in line or to whatever pattern you have selected.

Should you want to stop at any time before the job is completed, remove the mortar from the slab right up to the in-place flags. After wetting the slab down again or using a bonding agent, you can continue at any time.

FINISHING THE JOINTS

Wait several days until the mortar is thoroughly hard and you can walk on the flags without danger of loosening them. Mix up a little more of the same mortar, keeping it as dry as before. With a trowel sift mortar into the spaces between the flags until you have slightly more than filled them up. Then use the point of a trowel to compress the mortar in the joint. A caulking trowel, which is long and skinny, is best for this job, but with care you can make do with any pointed trowel. Slide the trowel along the filled joints to make their surfaces smooth and firm.

If you do this carefully you will not stain the flags. But should the mortar stain the flags, wipe the mortar off immediately with a wet rag or sponge. If the stain remains after the joint has hardened, use Brick Bath or any other cement-removing chemical cleaner. (*Follow instructions carefully; they all are powerful acids.*)

Some masons make a soupy mix consisting of 1 part cement to 3 parts sand and pour it into the joints using a partially flattened tin can to do the job. You may find this easier.

16

PIER FOUNDATIONS

Piers are masonry building supports. They are the easiest and the least expensive way we presently have of safely supporting a building or a porch above the ground. You can make piers from concrete alone, from concrete and concrete block, from concrete and brick, brick alone, or a combination of any masonry material. The choice of material doesn't matter very much. What is important is that you place the bottom of the pier below the frost line and make it large enough to safely support its load. This latter requirement is easily computed, and if you want to figure it out for yourself, the simple steps involved are covered at the end of this chapter.

LAYING OUT THE JOB

Examine the plan for the building. If it has been designed to be carried on piers, the plan will show the number of piers that have to be used, their top dimensions, and their spacings, which are generally given center to center. If the plan hasn't been prepared for piers, *don't start* installing them until the plan has been redrawn or otherwise changed to utilize piers. Some buildings can be carried by piers without major structural changes, others cannot, so let caution temper your enthusiasm.

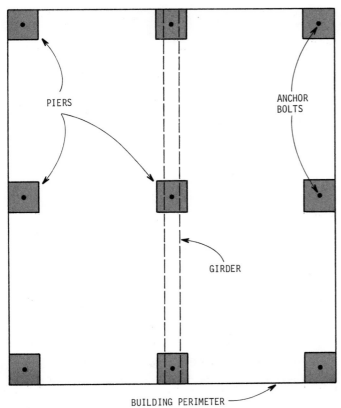

PIERS

ANCHOR
BOLTS

GIRDER

BUILDING PERIMETER

Piers are easiest and least expensive method of supporting a small building or porch safely above the ground. The usual positions of the piers beneath a rectangular building with a central girder are shown in the drawing at left.

219

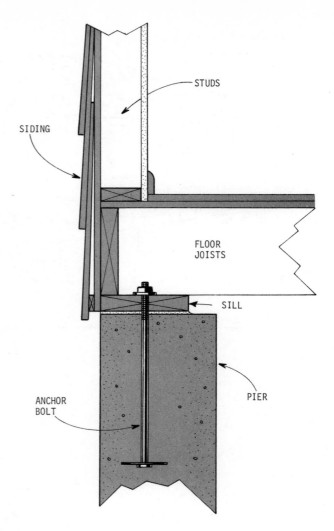

STUDS

SIDING

FLOOR
JOISTS

SILL

PIER

ANCHOR
BOLT

The frame of a house—that is, the floor joists—should be flush with the sides of the piers. To prevent rain water from collecting on top of the piers, the siding should extend below them.

Locating the piers. Let us assume that the cottage is to be a single-story, rectangular building 28 by 32 feet overall. In order to use comparatively short floor joists and thus save money, the most likely design would include a girder down the middle of the building. Let us further specify that this girder will be flush with the floor joists; therefore, the piers must all be level. Most likely the plan will call for nine piers, each 16 by 16 inches square at the top, with an anchor bolt embedded in its center. Thus the plan would show the centers of the corner piers 26 feet 8 inches and 30 feet 8 inches apart, making the outside corners of the building flush with the outside of the piers. (Actually, you want the building's siding to overlap the outside of the piers. Therefore in this case you would make the building's frame 28 by 32 feet.)

The first step is to locate the two front piers. Decide where the front of the building can be most suitably placed (and is acceptable to the municipality in which you may be building) and stretch a line to mark the exact front edge. Next, stretch two more lines to mark the exact sides of the building and one more to mark the rear.

The correct way to work with guide lines is to use batter boards. For a four-sided building you can use eight boards, each about 2 to 3 feet long, supported by a pair of sturdy stakes driven into the ground. The boards are placed about 8 feet away from the proposed building edges, and more or less facing each other at the corners of the building, as shown in the drawing. Stretch four lines from one board to another and thus mark the perimeter of the structure. Level the lines with a line level. Then raise or lower the batter boards as necessary. To determine if the corners formed by the lines are perfectly square, measure from corner to corner diagonally in both directions. If the distances are equal, the corners are square.

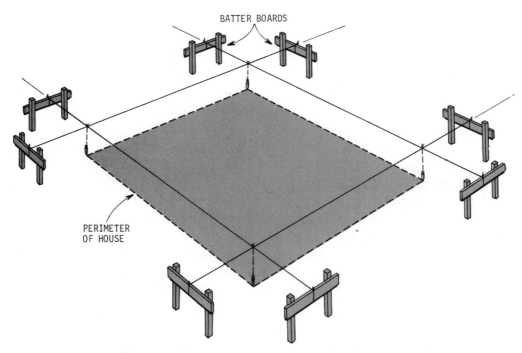

A typical layout with batter boards and lines for determining the perimeter and location of a building. To transfer the dimensions to the ground, drop four plumb bobs.

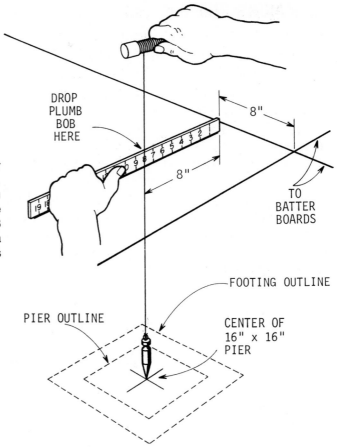

DROP
PLUMB
BOB
HERE

8"

8"

TO
BATTER
BOARDS

FOOTING OUTLINE

PIER OUTLINE

CENTER OF
16" x 16"
PIER

How to locate the center of the pier and the footing. If the piers are to be, say, 16 by 16 inches square, you must measure 8 inches from the corner (cross lines) and then 8 inches in from the line. Dropping a plumb bob from that point locates the center of the pier.

At this point you have four level lines marking the perimeter within which you are going to construct your building. If you build the piers with their outside edges flush with the lines, the outside edges of the building will be flush with the piers, which is of course what you want.

A building of this size will probably require eight piers evenly distributed around its perimeter with a single pier (or possibly two) under the center girder. The base of each will vary with the load-bearing capacity of the soil. Let us assume that in this case the soil is soft clay, and, according to the accompanying table, has a minimum load-bearing capacity of 1 ton per square foot. If you make your footings 3 feet square and 15 inches thick, this, as we shall compute farther along, will give you a considerable safety margin.

MAKING THE FOOTINGS

The footing is the lowest portion of the pier, the part that rests directly on the soil. The pier itself rests on the footing, more or less in its center.

The first step is to locate the centers of the holes you have to dig into the soil for the footings. To do this you refer to the blueprint to find out where the anchor bolts go in relation to the four lines you have stretched from the batter boards. Remember, these lines mark the exact outside edges of the piers and of the building you are planning to erect. Since the anchor bolts are in exact vertical alignment with the center of the footings, once you have located the position of an anchor bolt, you know where the hole for the footing must be dug.

LOAD-BEARING CAPACITY OF VARIOUS SOILS AND ROCK IN TONS PER SQUARE FOOT

	Approximate depth below grade	
Material	**3 feet**	**8 feet**
Soft silt and mud	0.1–0.2	0.2–0.5
Silt	1–2	1.5–2
Soft clay	1–1.5	1–1.5
Firm clay	2–2.5	2.5–3
Clay and sand	2–3	2.5–3.5
Fine sand	2	2–3
Coarse sand	3	3–4
Gravel and course sand	4–5	5–6
Soft rock	7–10	7–10
Bedrock	20–40 (average 25)	20–40

Start at any corner, where two lines cross. Measure along either line the distance specified on the print. Mark that spot. Then measure inwards, towards the center of the building, the required distance. Fasten a plumb bob (a pointed weight) to a string and hold the string end next to the relevant mark on the rule. When the bob stops swinging, its point is directly below the mark on the ruler. In this way you can transfer any desired point upwards or down.

With the center of the pier (and the footing it will rest on) marked on the ground, you can excavate the hole for the footing. The bottoms of the holes for the footings, in our example, should be 3 feet square and below the frost line. The sides should be parallel with, and outside, the batter-board lines.

Pouring the footings. The easy way is to simply pour a 1 : 3 : 5 mix of moderately wet concrete into the holes, in this example to a height of 15 inches. Use the end of a rake to make the surface of the poured concrete fairly level, but not smooth. It is best to pour the concrete as soon as you have excavated the holes to prevent water from collecting in the bottoms. However, if water does collect, use a minimum of water with the mix and pour gently. The weight of the concrete will drive the water out of the hole; even if much of the water remains, the concrete will harden beneath the water and little strength will be lost.

If the earth is firm, dig a straight-sided hole, below the frost line, as large as the required footing. The center of the hole should coincide with the center of the pier. Then pour concrete into the hole and let it find its own level. Make the footing at least 8 inches thick.

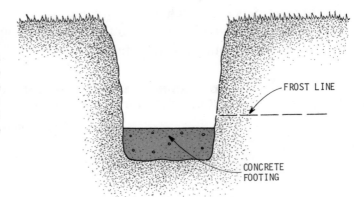

FROST LINE

CONCRETE FOOTING

When the earth is soft and a straight-sided hole cannot be dug, construct a square form, level it, and center it beneath the plumb bob which locates the center of the pier.

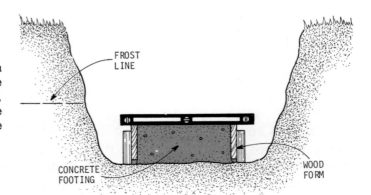

FROST LINE

CONCRETE FOOTING

WOOD FORM

CONSTRUCTING THE PIERS

You can make the piers out of concrete block, brick, cut stone, or poured concrete. You can also use precast concrete piers which you simply lower into place. The following instructions explain how to make a pier from poured concrete.

At this point you have poured all the footings for the piers, and there are four lines marking the perimeter of the building you are going to erect. Now you have to locate the piers on the footings.

Starting at a corner, drop a line from the point where the two lines cross. Mark this point on the concrete footing. This is the outside corner of the corner pier. Measure 16 inches along one of the lines from the same corner and mark this point on the concrete by dropping a line. Detach the plumb bob and repeat the measuring and marking operation 16 inches from the corner along the other line. You now have three marks on the surface of your footing indicating the corner of the pier you will erect. All you need do now is erect a perfectly vertical pier within these three marks.

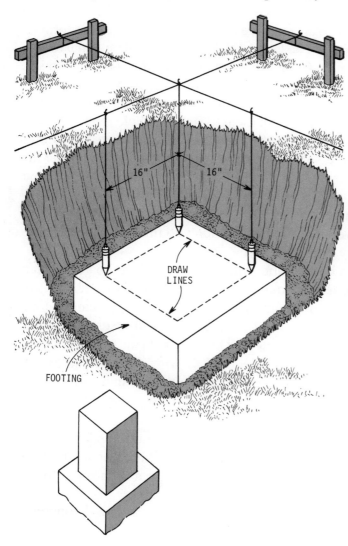

16" 16"

DRAW
LINES

FOOTING

To lay out the position of a pier on a footing, three lines are dropped, as shown, or one plumb bob dropped three times. One line is dropped from the crossed lines; the other two 16 inches from the corner. Small drawing shows how pier will look when finished.

The same method is used to locate the other corner piers. To locate the piers along the sides, you simply measure along a side line the required distance, drop a line, then measure another 16 inches and drop a second line. This gives you two marks on the footing that coincide perfectly with the side of the projected building. Draw a line between these two points and then use a square to lay out the rest of the area on which the pier will rest.

Cast-concrete piers. Construct a long square form of 1-inch lumber, open at both ends, with a cross section equal to that of the desired pier. Oil the inside of the form. Since there will be considerable outward pressure on the lower end of the form, reinforce the joint by wrapping heavy galvanized wire around the outside of the form or nail strips of metal around the corners. The length of the form should be equal to or a fraction less than the height of the pier. Center the form over the footing. With stakes and braces "lock" the form in place so that it will not move or rise when you pour concrete inside. One way to do this is to position the form and then nail a pair of 2x4s to its sides. Place rocks on the 2x4s or nail them to stakes to hold them in place.

Using the same mixture suggested for the footings, pour the concrete gently into the form. Then poke a 2x4 into the soft concrete to make certain there are no air pockets. Insert the anchor bolts as indicated on the blueprint. Give the poured concrete a day or two to harden, then remove the form and use it again.

If you pour each form an inch or two short of what you need, you can easily correct its height later by adding a layer of mortar. If you don't do this you have to make certain the form's height is exactly right before you pour.

Poured piers are definitely the best solution when the footings are barely wider than the pier itself and the pier doesn't extend more than a foot or so above the ground. In such cases dig the holes with their sides as vertical as you can and no larger than the footing cross section. Construct the pier form and insert it into the hole with its top exactly at the desired pier height. Then place a pair of 2x4s across the hole and nail them to the form. Hold the 2x4s in place with rocks or stakes. At this point you have a form extending partway into a hole in the ground, its top edge at the correct height and in exactly the right position in relation to the rest of the piers.

Use the concrete mix suggested. Pour slowly and carefully until the level of the concrete reaches the bottom edges of the form. Then wait a half hour or so until the concrete stiffens somewhat. Then you can pour the rest of the mix into the form, strike it flush with the top and insert the bolt.

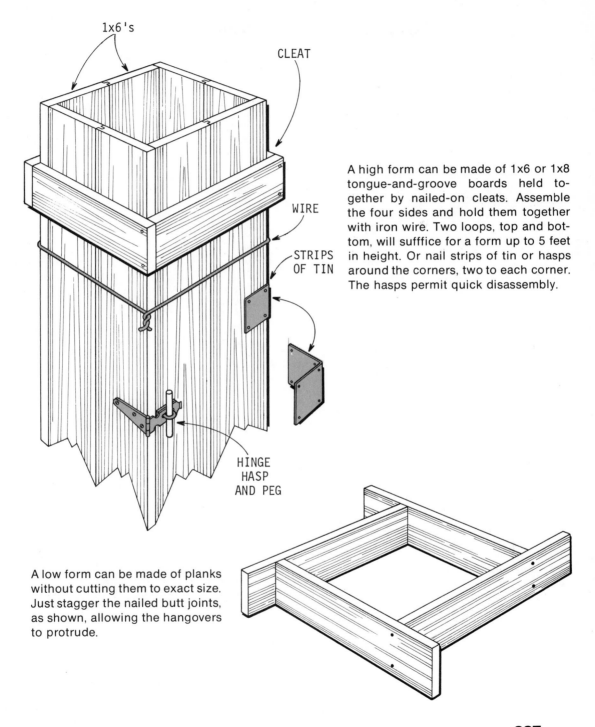

1x6's

CLEAT

WIRE

STRIPS
OF TIN

HINGE
HASP
AND PEG

A high form can be made of 1x6 or 1x8 tongue-and-groove boards held together by nailed-on cleats. Assemble the four sides and hold them together with iron wire. Two loops, top and bottom, will sufffice for a form up to 5 feet in height. Or nail strips of tin or hasps around the corners, two to each corner. The hasps permit quick disassembly.

A low form can be made of planks without cutting them to exact size. Just stagger the nailed butt joints, as shown, allowing the hangovers to protrude.

THREE WAYS TO POUR A FOOTING AND PIER

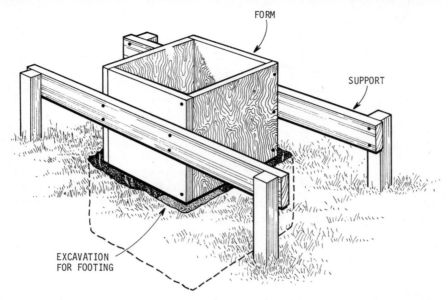

FORM

SUPPORT

EXCAVATION
FOR FOOTING

1 When the footing does not have to be large or deep, and the pier is fairly low, dig a hole that is slightly larger than the pier form and brace the form as shown. Then pour footing and pier together.

2 When the ground is hard, the bottom of the pier can serve as its own footing. Allow the form to reach the bottom of the hole and make one pour. Pour slowly so as not to burst the bottom of the form.

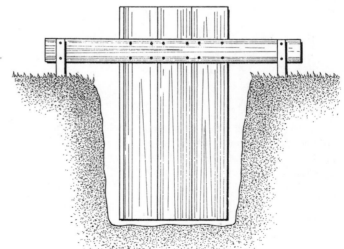

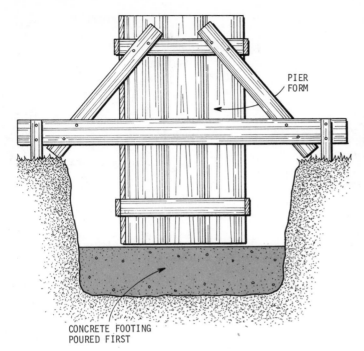

PIER
FORM

CONCRETE FOOTING
POURED FIRST

3 When the footing has to be larger than the pier, lift and brace the pier form as shown. Pour the footing to the level of the bottom of the pier form. Wait until the concrete sets up and then pour the pier. Or, pour the footing; then brace and level the pier form atop the footing and pour the pier.

Leveling the piers. Construct one pier and let it serve as a guide to determining the height of the others. There are several devices that will help you to level the piers. The easiest to use is a transit or sighting level. It can be rented; no need to buy one for just one job. The sighting level produces the most accurate results as well as being the most convenient tool to use for leveling.

Set the sighting level up somewhere near the middle of the perimeter formed by the lines. Adjust the screws until the bubbles are centered when you swing the sight all the way around. When this is so the sight is perfectly level. Point the sight towards the finished pier. If you see the top edge of the pier in line with the horizontal hair, all is fine. If the surface of the pier is below your sight, have a helper stand a ruler vertically on the pier. Read the number that is visible at the cross hairs. Let us assume you see 10 inches. This means your line of sight is exactly 10 inches higher than the top of the pier. Swing the sight round to any other pier that you are constructing. Have your helper hold a ruler vertically on the footing. Read the relevant figure. Let's say it reads 60 inches. Subtracting the first figure, 10, you get 50, which means the pier at that point must be 50 inches high to be level with the first pier. As you erect the second pier, resight and add mortar, if necessary, to level the pier.

HOW TO USE A TRANSIT

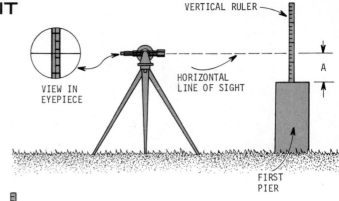

1 The transit is set up in the center of the building's perimeter. The leveling screws are adjusted until the bubbles in the spirit levels are perfectly centered. The first pier is centered in the transit's scope. If the line of sight is above the top of the pier, as shown in the illustration, a rule is placed on top of the pier. The height of the line of sight (*A*) is read on the rule.

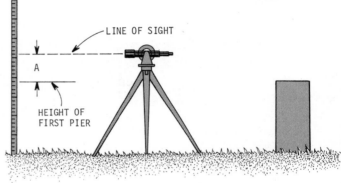

2 The scope is turned until it points directly over the location of another pier (any pier), and is sighted directly on a rule held vertically at that point. The top of the pier at this location will be the sighted reading on the rule, less dimension *A*. Now the scope can be turned to sight another pier.

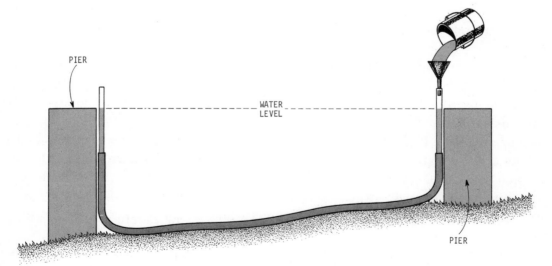

A water level is an aid in making all the piers the same height. The water in both glass tubes will always be the same height. You can make one of hose and plastic tubes.

230

If you cannot rent a transit, you can use a water level for the same purpose. At this writing the device with 50 feet of tubing sells for less than $12. It consists of a hose with two clear plastic tubes connected to its ends. The hose is filled with water, and obviously, since water seeks its own level, the water at one end of the hose is always at the same level as the water at the other end. To use this level, have a helper hold one clear tube next to the finished pier with a few inches of tubing above the finish line. Carry the other end of the tube to the pier under construction. Lift the clear section up until you can see water. If necessary raise or lower both tubes until the water level in the tube next to the finished pier is level with the top of the pier. The water tube is not the most convenient device for leveling. but it is highly accurate.

Still another alternative, less accurate but far cheaper, is the lowly line level, which is a short level that hooks onto a line. To use it, stretch a line from the top of the finished pier to the second pier. Pull the line taut and move the level to the middle of it. Lift or lower either end of the line until the bubble is centered, at which point the two ends of the line are level with one another to within about 1 inch in 50 feet, which is more than accurate enough for a small house.

There are also hand sighting levels for under $10 and sights that clamp on to the ends of a spirit level. You can use both of these devices for leveling, but they are only really useful for preliminary work.

WORKING ON HILLS AND ROCK

Hills present certain problems. The piers on the low side of the hill must be level with those above. You must make certain that the footings are below the frost line and that the bottoms of the excavations for the piers are level.

Should you encounter perfectly flat bedrock, you can eliminate the footing. Bedrock can carry 25 tons per square foot. Just build the pier right on top of the stone after you have carefully cleaned it.

Sloping bedrock is another matter. No matter how carefully you clean or roughen it, or even if you use a bonding agent, there is a good chance your piers (and house) will slide down the hill. In order to prevent this with certainty, use a star drill or an air hammer and drill two or three $3/4$-inch holes 6 inches deep in the rock beneath each pier. Insert 1-foot pieces of $3/4$-inch steel reinforcing bars into each hole.

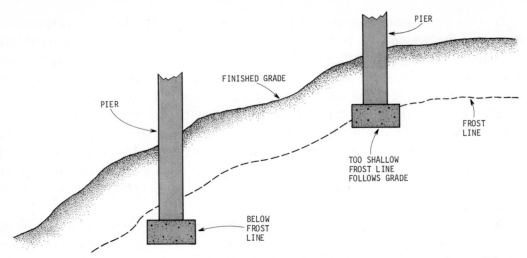

When digging a hole for a footing on the side of a hill, make certain the bottom of the hole is horizontal and below the frost line.

There is no need for a footing when building on bedrock, unless the rock is pitched. Then you must drill holes into the rock and lock the footing in place with steel bars.

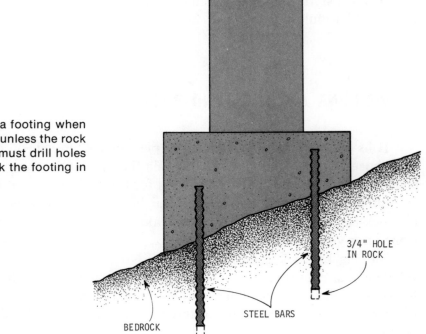

Then pour the footings or lay the block around the bars, filling the space between the block and the steel with concrete. The bars will keep the footing from sliding downhill.

CALCULATING FOOTING DIMENSIONS

If you want to determine whether footings are large enough to preclude any danger of their sinking into the earth, here is the basic calculation. It is simple enough.

Figure the roof at 35 pounds per square foot, dead and live load. Thus, if the roof is 28 by 32 feet, as in our example, you have 896 square feet of roof, which multiplied by 35 gives you a roof load of 31,360 pounds.

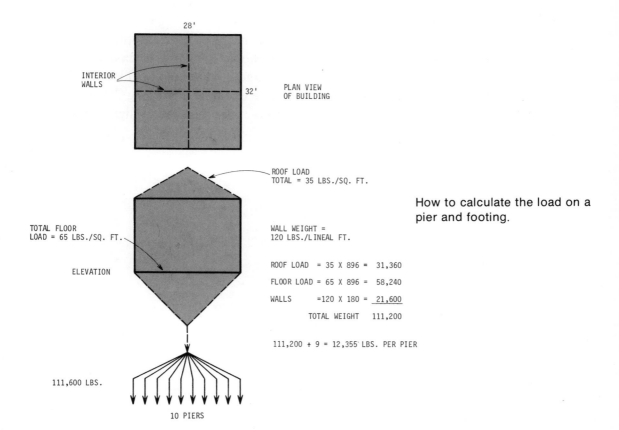

How to calculate the load on a pier and footing.

First and second floors are rated at 65 pounds per square foot for both live and dead loads. So in this single-story example, you would have a load of 896 by 65, or 58,240 pounds.

In addition you have walls. Standard walls are generally figured at an average weight of 120 pounds per linear foot. The perimeter of the building comprises 120 feet of wall, and if you add another 60 feet of interior wall, you have a total of 180 feet of wall, which at 120 pounds per foot multiplies out to 21,600 pounds.

The total weight of the roof, one floor, and walls is therefore 111,200 pounds. You then divide this figure by the number of piers to be used. Supposing there are 9, this gives a pier load of 12,355 pounds per pier and footing. Dividing this by 2,000 gives you 6.2 tons per pier and footing. This load is well within the capabilities of concrete, concrete block, and brick and stone. The limiting factor is the load-bearing capacity of the earth. Soft clay, which has a low rating, can carry 1 ton per square foot. Having made the footings 3 by 3 feet, you have 9 square feet of supporting surface, which is capable of carrying 9 tons. This gives a good safety margin.

The relation between the pier and its footing is simple enough to establish by rule of thumb. Never make the pier less than half the size of the footing. For example, if a footing measures 2 by 3 feet, the pier should measure at least 1 by 1½ feet. Never make a footing less than 8 inches thick and never make its thickness less than about 40 percent of its largest dimension. For example, a footing 3 feet wide should be about 15 inches thick. If you require piers more than 5 feet high, get local engineering advice.

17

PERIMETER FOOTINGS

While the construction of a permanent, year-round home may appear to be a formidable project far beyond the capabilities of an individual not familiar with the construction trades, this is in fact not so. Thousands of homes are presently being constructed by their owners all over the country, and some of these people have never held a hammer in their hands before, let alone read a building plan. The reason for this effort is mainly economic. The average construction worker costs the home purchaser close to $15 an hour (base salary, insurance, fringe benefits, contractor's profit, etc.).

In my neighborhood new homes, actually cottages by yesterday's standards, sell for around $65,000. The breakdown is approximately this:

$15,000	building lot
3,600	broker
1,400	legal and mortgage costs
20,000	materials
20,000	labor
5,000	builder's profit
65,000.	

Thus, if you were to build your own home in my neighborhood you could cut your costs by approximately $28,600, which means a 44 percent saving. In other areas, where the price of the building lot is less exorbitant, say $5,000, you could build a home for around $21,400 instead of $55,000, which is a considerable saving. We have assumed in our example that you work from standard plans and do not hire an architect, which would cost you an additional 5 to 10 percent, depending on the size of his office.

One very important point to remember at all times when building your own house is: *build to a minimum*. Don't stint on size, because next to location, that is the most important feature of a house, but do not overspend on materials. I can only tell you of an acquaintance of mine who ran up a $34,000 bill constructing a four-room cottage in the days when such cottages sold for $14,000 complete. Not only did he exhaust his life's savings, he had to sell the place without ever moving in.

The construction of a permanent home begins with either a slab foundation, discussed in the next chapter, or a perimeter footing, covered in this chapter. While it would be misleading to say that the construction of either footing is easy — it is not, a lot of physical work is involved — it would be equally inaccurate to say it is technically difficult. This may be hard to accept as you stand there looking at your building lot, which may be covered with bushes and trees, and wonder how you can be certain you will position the footing properly and keep it square and level. But in fact, this isn't difficult at all. It can be and is regularly done with no more than a line, steel tape, plumb bob, and line level. The degree of accuracy required is modest. Anything within an inch, plus or minus, will do at the top of the foundation wall that will be erected on the footing. So even if your footing pitches 2 or 3 inches, there is nothing to worry about, since you can easily correct this by the time you reach the last course of blocks. The same is true of out-of-square corners; they can be pulled and corrected without much difficulty before you reach the top course. As for the position of the building itself on the lot, no one is going to complain if your home is a couple of inches from where the municipal code requires it to be.

As for the physical labor involved, 95 percent of the work can be done by machine. The rest is routinely accomplished by a mason and a helper on many jobs (with a few extra hands for cellar floors). The average 30- by 40-foot footing and foundation wall can be erected by two men in two or three weeks, depending on the complexity of the job.

Another point to keep in mind is that you can always find contractors willing to pick up the job at any stage if for any reason you wish to stop working.

LAYOUT PRELIMINARIES

Let us assume that you are going to construct a perimeter footing for a building 30 by 40 feet on a fairly level lot. This means that the exterior dimensions at the bottom of the structure, where it rests on the foundation wall, will be 30 by 40 feet. Of course, all footings are not simple rectangles, but even the more complicated ones are constructed in the same way.

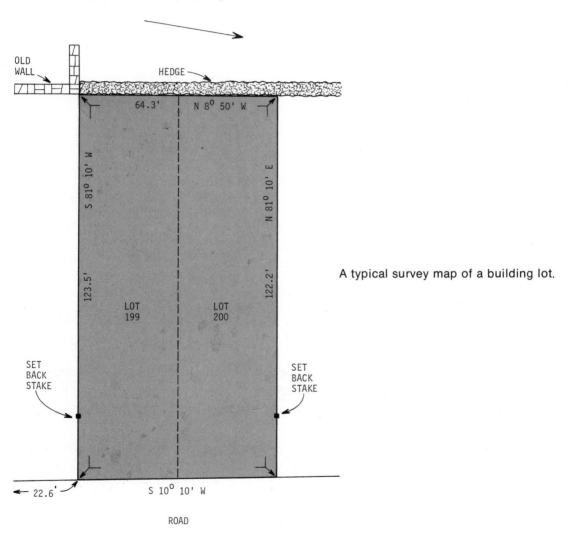

A typical survey map of a building lot.

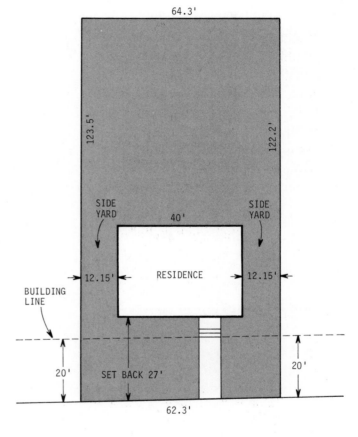

A typical foundation plan. Dimensions of the set back and the side yard must comply with your code.

Although this book is not primarily a guide to home construction, the steps necessary before beginning excavation are included for your convenience. They are the most important of all, assuming you are constructing within a municipality.

Surveying the land. Unless you already have a plot plan drawn by a licensed surveyor, you must by local law hire a licensed surveyor to survey the property, mark it with stakes and furnish you with a signed map. This must be done even if your family has always lived on the property and you know its borders exactly. You will need this map when you submit your house plans to the building department for approval and when you seek a mortgage.

238

Be sure to ask the surveyor to lay out the setback, which is the line that determines the front edge of your building in relation to the street. In some localities you must hold to this line. In some, you can position the building farther back if you wish.

Building department requirements. The building department must approve of your building as such and must also approve of it in relation to your lot.

Most building codes prohibit look-alikes, which means your home cannot resemble its neighbors too closely. Many limit the minimum and maximum size of the building, and they all have setback and side-yard rules that establish the minimum permissible space between your house and your neighbor's property line.

Among other requirements, the building department will also specify minimum foundation wall thickness and minimum footing width and thickness. Be advised that this minimum is probably the maximum ever used in other countries. This is why our buildings never fall down by themselves—a fairly common occurrence overseas. There is therefore no need to increase the building department's more than generous minimum dimensions.

EXCAVATING PRELIMINARIES

Demarking the excavation. At this point the lot has been surveyed, and the plot and building plans have been submitted to and approved by the building department. Your next steps are to remove whatever brush and trees may interfere with the work and to mark the perimeter of the excavation.

Start by locating the six stakes driven into the ground by the surveyor. Stand with your back to the road, facing the front of the projected house. Let us assume there are three surveyor's stakes demarking the left side of the property. Find them and stretch a line directly above them, using two of your own stakes to hold the line, and a plumb bob to make certain the line is directly over the carpet tacks driven into the tops of the surveyor's stakes. This line now accurately marks the left side of the property.

Assuming the left side yard is to be 10 feet wide, measure inwards from the left property line a distance of 10 feet at two places along the line and mark the points with a couple of stakes. Repeat the procedure on the right side of the house, following the measurements specified in the plan.

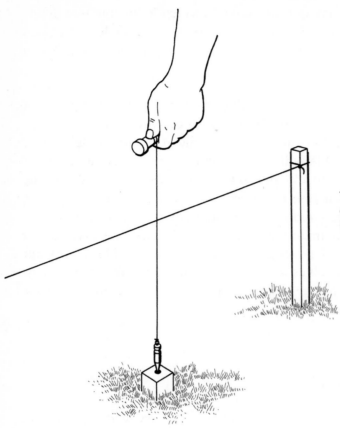

To make certain the lines are directly over the survey stakes, drop a plumb bob at each corner.

Next, locate the middle surveyor's stakes on both left and right property lines and stretch a line between them. This line marks the front edge of the house. Measure rearwards from this line and stretch another, parallel line marking the rear edge of the house. If the front to back dimension of your building is to be 30 feet, the two lines must be exactly this distance apart.

Installing batter boards. As previously explained, batter boards are 1x8 boards 3 to 4 feet long, supported by two stout stakes. You need eight batter boards and accompanying stakes.

Install a set of batter boards about 10 feet from each front corner of the house. Transfer the line marking the front of the house to the boards. Drop a bob from the line to each surveyor's stake, to center the line directly over the stakes. Hang a level

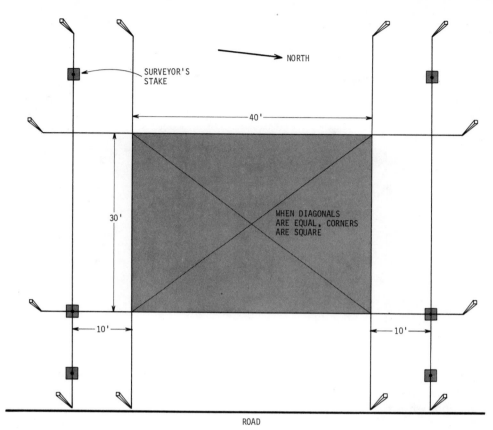

NORTH

SURVEYOR'S
STAKE

40'

30'

WHEN DIAGONALS
ARE EQUAL, CORNERS
ARE SQUARE

10'

10'

ROAD

Layout of the foundation of a building. Measure carefully and move the building lines
as required, making sure the corners are square.

in the middle of the line and raise or lower the batter boards as required. Install batter boards at the rear corners also.

There are stakes roughly marking the left edge of the building. Use these to guide you while installing another pair of batter boards about 10 feet beyond the front and back lines. Stretch a line over these boards and position it accurately by measuring from the left-side property line. Hold the tape horizontal and transfer measurements vertically with the aid of a plumb bob. Repeat the procedure at the right side. Then level all four lines. Remember, the level must be in the center of the line to be accurate.

You now have lines marking the right and left sides of your property and four lines demarking the perimeter of the house you are going to build. Remeasure every-

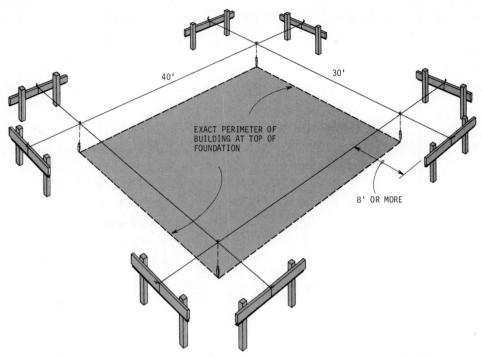

EXACT PERIMETER OF
BUILDING AT TOP OF
FOUNDATION

8' OR MORE

When the building lines have been transferred to the batter boards the situation looks like this. Keep the batter boards far enough away from the prospective foundation to avoid interference.

thing to be certain. Check for squareness by measuring the corner-to-corner diagonals. All is square if these two distances are equal.

Next, carefully mark the position of each line on the batter boards. You are going to remove these lines and you don't want to have to go through the entire measuring process again.

This done, take some powdered lime and trace a box two feet beyond the lines that mark the perimeter of the house. The box, which will thus be 4 feet longer and wider than the house foundation, will provide preliminary guidance for the bulldozer. Now remove the lines.

Excavation depth. Before digging you have to decide just how deeply to dig. There are a number of interacting conditions that must be satisfied when you are deciding on excavation depth. At a minimum you must go below the topsoil to reach firm subsoil. In frost country, you must continue down below the frost line. For a full cellar and a better-looking home you should go down deep enough to leave no more than about 2 feet of masonry projecting above the original grade. This is a desirable height because it permits you to slope the soil upwards to the building, speeding rainwater runoff, and still leaves a foot of concrete or block above grade as a termite barrier.

At the same time you must keep the municipal sewer in mind, if there is one. The floor of your toilet must be sufficiently above the sewer main to permit a downward pitch of ¼ inch to the foot or a little more.

A word about full cellars. Assuming that you eventually want to "finish" the cellar, you will need fourteen courses of concrete block to provide an interior height of a little more than 9 feet. This is necessary because you will need at least 1 foot below the ceiling in which to hide the pipes to end up with a finished cellar height of 8 feet.

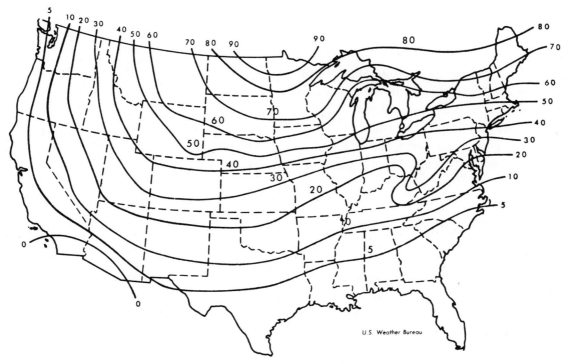

Depth of frost in various parts of the country.

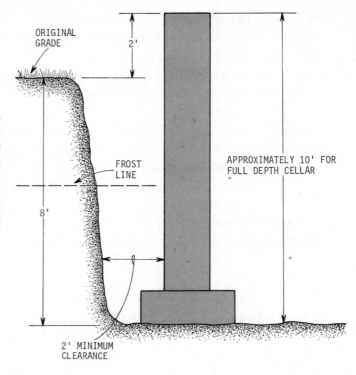

A perimeter footing should go below the frost line and there should be sufficient space between the side of the excavation and the footing to permit working. Building the foundation 2 feet above grade permits you to slope the earth up the foundation for water runoff and still leave a foot for termite protection.

ORIGINAL GRADE

2'

FROST LINE

APPROXIMATELY 10' FOR FULL DEPTH CELLAR

8'

2' MINIMUM CLEARANCE

The exterior, overall height of the footings and foundation wall will be about 10 feet when the foundation wall itself is about 9 feet high. Since you want 2 feet of foundation wall exposed, it is necessary to excavate to a depth of 8 feet. The sewer main still has to be taken into consideration. If the main is low enough to permit the pitch already mentioned, fine. If it is higher, you may have to reduce the depth to which you dig or make the foundation wall higher.

EXCAVATING

Excavating contractors work by the day or the job. If the job is simple, no soil has to be hauled off and there are no large tree stumps to be pulled, you can probably do financially best on a day-hire basis, bearing in mind that a day-price is not cheap if the machine is small. (Bulldozers are rated by blade size.) If the job involves stump and soil removal, it is best to set a price for the entire job, including backfilling and rough grading. In either case, make it your business to be there when the machine is working. Remember, what the machine does not move you will end up moving by hand.

Holding depth. It is important that the excavation be held as close as possible to the desired depth and also that it be kept as level as possible. This is almost impossible for the machine operator to do himself, because he has only his eye to guide him. He needs help.

Several methods can be used. The easiest requires two men and a transit. After preliminary excavation has been completed, the transit is set up to one side of the job. It is pointed over the original grade at a ruler held vertically in some spot there. Assume the sight in the transit level shows 2 feet at the cross hairs. This means the transit is 2 feet above grade. Let us suppose you want to go down 8 feet. A long ruler or marked stick is then held in a vertical position at a chosen spot in the excavation, and the instrument is sighted at it. If the 10-foot mark appears in the cross hairs, the bottom of the excavation is exactly 8 feet below grade. If the mark is higher, you have to dig deeper. The stick is positioned at various places in the excavation until the levelness and depth of the entire excavation have been checked.

Another method requires that the four lines marking the perimeter of the house be reinstalled. The distance of the lines above the original grade is measured and then added to the desired 8 feet. When this total coincides with the distance from the bottom of the excavation up to the lines, the bottom is at the correct depth.

Still another method consists of stretching a line from the top of one side of the excavation to the other after a good portion of the soil has been removed. Use a tall stick with a mark to measure the depth.

A few excavating tips. Have the dozer push the topsoil to one side before you start really digging. You can use this soil later for landscaping. Be careful not to let the contractor cart away more soil than necessary. A shortage will cost you later, whereas extra can do no harm. At worst, you can always "hide" it. For example, spread a 2-inch-thick layer of soil over a lot 100 feet square and you have hidden 60 cubic yards of soil. Have the dozer move the excavated soil far enough away from the job so that you don't have to climb hills to work. And have a ramp dug into the side of the excavation. You will need this to bring material down into the center of the job. Be certain sufficient soil is left on hand to fill in this ramp and backfill around the foundation.

At this point you have a flat-bottomed trench, two feet wider on each side than the perimeter of the house. You need this clearance in part for the footing and in part as working space.

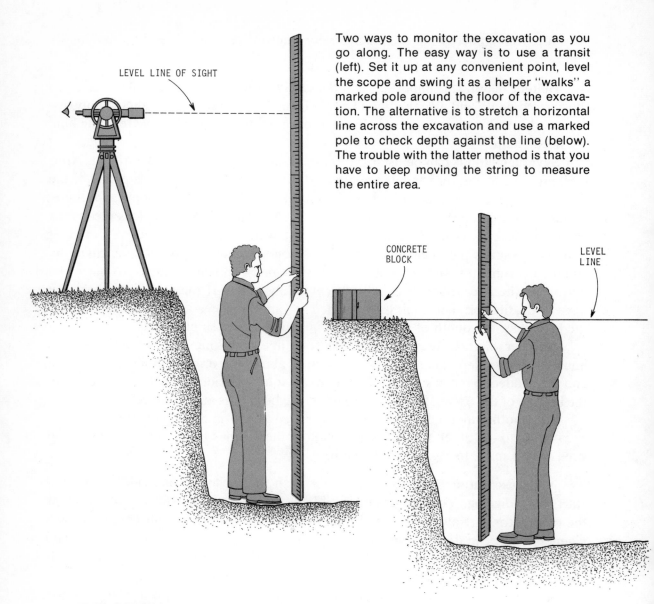

LEVEL LINE OF SIGHT

Two ways to monitor the excavation as you go along. The easy way is to use a transit (left). Set it up at any convenient point, level the scope and swing it as a helper "walks" a marked pole around the floor of the excavation. The alternative is to stretch a horizontal line across the excavation and use a marked pole to check depth against the line (below). The trouble with the latter method is that you have to keep moving the string to measure the entire area.

CONCRETE BLOCK

LEVEL LINE

THE FORMS

Footing cross section. Eight- or 10-inch concrete block is usually specified by the local building code for use in constructing two- and three-story structures. By rule of thumb, the footing should be twice as wide as the block. Therefore, if you are going to use 10-inch block, the footing should be 20 inches wide. Thickness should be half

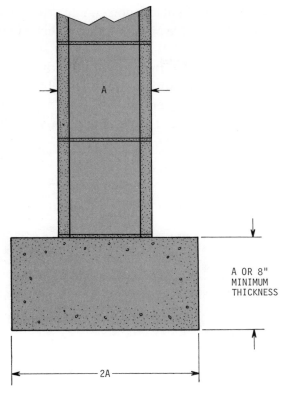

A OR 8"
MINIMUM
THICKNESS

2A

Generally accepted rule of thumb for determining footing dimensions. Footing width should be at least twice the width of the block or poured concrete foundation wall. Footing thickness should be at least equal to width of block, but never less than 8 inches.

the width. Thus a 20-inch wide footing should be 10 inches thick. This is generous for almost all soils and loads. However, to be quite sure you can figure out the specifications following the suggestions given in the preceding chapter. Obviously, the entire perimeter serves as a support. Thus a 30- by 40-foot footing 20 inches wide would have 33,600 square inches of concrete on the ground. Dividing by 144 gives 233.3 square feet. On soft clay, this footing could support 233.3 tons, since clay can carry 1 ton per square foot.

Constructing and positioning the forms. Make forms of 2-inch planks of suitable width (8-inch for 8-inch concrete, etc.). Place the planks on edge in the trench and support them with wood braces nailed across their edges every 10 feet or so.

The next step is to position the form. Find the marks you made on the batter boards. You removed the lines in order to dig the trench; now replace them on the marks. If a 10-inch block is centered on a 20-inch-wide footing, there will be 5 inches of footing on either side of the block. Thus you must move the line 5 inches *outward* to mark the edge of the footing. With all the lines in place, you can drop a plumb line at any point and accurately establish the outside edge of the footing. (This will be the inside surface of the form board.)

Drive stakes into the ground along the outside of the form to keep the form from moving from side to side. Next you must make your form perfectly level. This can be done with the aid of a transit, water level, or line level. Do not use a spirit level as even a slight misreading of the bubble will lead to a considerable error in 40 feet of form.

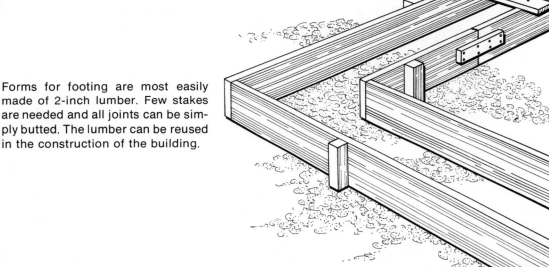

Forms for footing are most easily made of 2-inch lumber. Few stakes are needed and all joints can be simply butted. The lumber can be reused in the construction of the building.

Raise or lower the form as required, digging a little or using stones to raise it where necessary. When the form is perfectly level, nail it to the stakes. You don't have to cut them flush.

If you find large hollow areas beneath the form where concrete may escape, bring these areas level with crushed stone or gravel.

Reinforcement. If the soil is very soft and moist, you may want to reinforce the footings to reduce the chance of their cracking. Use three or four 1/2-inch-thick, steel

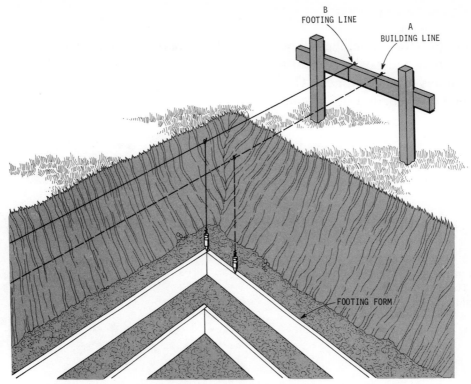

Use the batter board lines to locate the footing forms rather than measuring each time you drop the plumb bob. Merely move the line from position *A* to *B*.

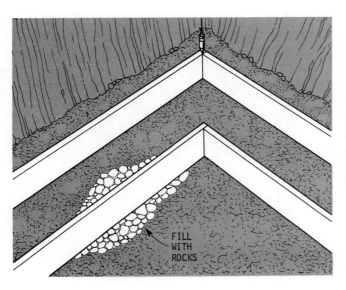

Fill all hollows beneath the leveled form with rocks. Do not use soil. In some localities the building inspector demands that he see the in-place forms before you pour.

reinforcing bars around the full perimeter of the footing. Space the bars more or less evenly apart and position them about 3 inches from the bottom of the form. Use small stones to support them. Overlap the bars about 6 inches or more where necessary and bend them to go around corners.

Installing a footing form in an excavation. The vertical line marks the other side of the form. The workman is checking the width with a rule.

Footing form is held in place by stakes and cross braces.

250

In this typical "pour" setup, a ready-mix truck is backed into position. Its chute is above a second metal chute. The driver watched the men for the signal to start pouring the concrete.

Inspection. In some cities and towns the local building inspector insists he see the footings and supporting soil before you pour. If this is the case or if you can get him down there even if it is not, this is highly advisable. You are paying his salary, so why not use his experience?

Lally columns. On a 40-foot-long building such as the one in question, the usual design calls for a girder down the middle supported by three Lally columns. Each column must rest on a concrete slab base. Each base can be 2 by 2 feet in size and 10 inches thick. The blueprint will show where the columns go, and since they need be no more than approximately centered on their slab bases, it is no problem to position the forms. But be certain to make the support forms flush or slightly below the top of the footing forms. Use any of the leveling techniques previously mentioned.

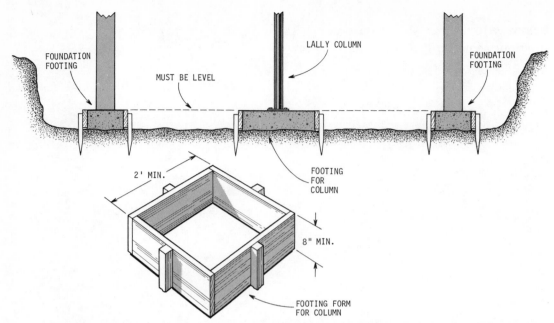

FOUNDATION
FOOTING

MUST BE LEVEL

LALLY COLUMN

FOUNDATION
FOOTING

FOOTING
FOR
COLUMN

2' MIN.

8" MIN.

FOOTING FORM
FOR COLUMN

Footing for lally column should be as thick as the footing supporting the foundation and at least 2 feet square. Its top surface should be level with the top of the footing, or an inch or so lower, but never higher.

When mixing your own concrete, you can pour at your convenience. If you are using ready-mix, it is easier to pour when the truck is on hand. However, if the truck has to enter the area between the footing forms and you have to move the Lally forms, it is best to reposition these forms after the truck has gone and fill them by hand. The reason is that the heavy transit-mix truck may well compress the soil so much that your Lally forms won't be level if you simply move them back into place.

Filling the forms. Get yourself and your crew ready. See that each man and woman has a rake or hoe. When working with ready-mix, check to make certain the truck's chute can reach all parts of the form (or far enough to allow the concrete to be pushed to all parts of the form). If this is not possible, consider removing a portion of the form to allow the truck to back into the center of the excavation. (This is one reason why you need that ramp.) Or construct a wooden chute to use with the truck's chute, or rent an extension chute from the concrete supplier.

If you mix your own concrete you don't have the problem of the truck's chute reaching the form, but you have another problem—that of pouring in sections. The best way to do this is not to use a smooth stop across the form, but to let the edge of the pour remain rough and uneven, so that when you continue later on, this surface will provide a better bond.

Unless the code specifies otherwise, use a 1 : 3 : 5 mix with sufficient water to permit you and your crew to move the concrete easily. Push the concrete along to the far end of the form and fill one section completely to the top before starting on the next. Then with a straight board screed the concrete level along the top of the form. When the form is completely filled and leveled, the pouring operation is complete.

Form removal. Wait a day or two until the concrete is fairly hard, then remove the form carefully. The sooner you remove the boards, the easier it is to do so, but the easier it also is to damage the concrete. If you wait too long, the boards become locked in place and you need a pickax to remove them. Unless you are in desert country there is no need to cure the footings. They will soak up all the water they need from the ground. Where the earth is dry and there is no rain, give the footings a good soaking every few days for a week or so.

"Pulling" the mix. Since the concrete has a long way to go to fill the entire form, a soupy mix is called for, and rakes and hoes are used to pull the concrete along.

253

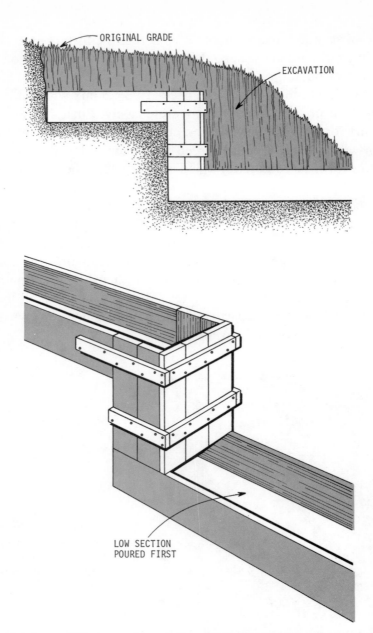

ORIGINAL GRADE

EXCAVATION

LOW SECTION
POURED FIRST

Pouring a footing on hilly ground requires a stepped form. Side view (top) shows relation of horizontal and vertical parts of the form to the surface of the hill. Construction details minus stakes and braces, are shown in bottom drawing. The lower portion of the form is poured first and permitted to set up; then the upper section is poured.

254

STEPPED FOOTINGS

If you find the beautiful lot you purchased sight unseen is on the side of a hill, or if you are not able to afford anything better than a hillside lot, do not despair. You can still construct your home on it by means of a stepped footing.

You drop your footing in a series of steps more or less following the slope of the ground. Each section of footing is perfectly horizontal and as far below grade as soil conditions and frost require. The sides of the steps that rise vertically from one horizontal section to the next are constructed by using an enclosed, vertical box form similar to those described for use as pier forms. The lower, horizontal section of the footing is poured first. When this has set up sufficiently, the vertical section is poured. Later the form is removed.

18

SLAB FOUNDATIONS

The slab foundation is simple and inexpensive. It eliminates the need for and expense of excavation, footings, and foundation walls and provides a floor at less cost than one made of wood. The slab foundation also eliminates the cellar and with it all cellar water problems. Supposedly a cellar is the least-used space in a home. However, without a cellar you have to give up above-ground space to storage, washroom or laundry room, and furnace. In addition, eliminating the cellar will often reduce the maximum possible mortgage by 25 percent or more. This is because mortgage lenders usually measure the value of a home by its cubic space, and in such computation, the cellar accounts for almost as much as the rest of the building.

In any event, in its simplest and least expensive form, which is suited to hard, dry soils such as are often found in our Southwest, the slab foundation is little more than a slab of concrete with a thickened edge. Where there is no danger of frost, but where the soil is not hard and dry, the slab rests on a shallow perimeter footing, which can be cast along with the rest of the slab. Up north, where frost is a problem, the foundation is poured in two sections. One is a perimeter footing that goes down below the frost line, and the other is a slab insulated from the ground and the perimeter footing.

All three types of slab foundations are covered here, the simplest first.

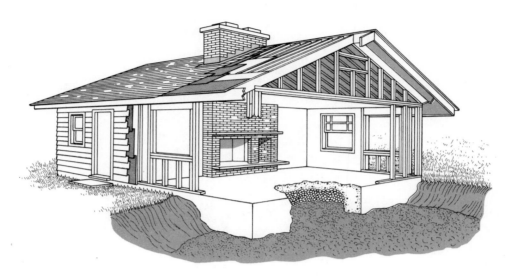

A house constructed on a slab foundation.

DESERT CONSTRUCTION

Where the soil is hard and dry and rain is something you read about in news-papers, the simple, thickened-edge, slab foundation may be used.

Laying out and excavating. Start by installing eight batter boards and lines to mark the sides of the slab using the same methods suggested for demarking perime-ter footings.

Remove the topsoil, if there is any, and simply level the soil for a distance of about 5 feet beyond the edge of the slab you plan to lay down. For example, if you are going to make a 30- by 40-foot foundation, clear and level an area 35 by 45 feet. Work from the lines you have stretched across your batter boards to hold the soil fairly level. Additional, final leveling can be done more easily later on.

Next, dig a trench 1 foot wide and 2 feet deep with its outside edge just under the lines marking the outside edges of the desired slab.

The footing (that is what it actually will be) that results when you fill the trench with concrete can easily carry a one-story frame house. If you want to construct a higher or heavier building (brick or stone) follow the suggestions given in Chapters 16 and 17 for computing required footage dimensions on various soils.

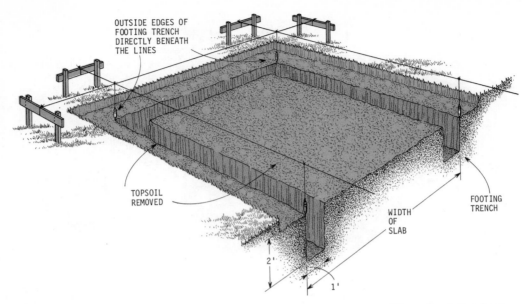

OUTSIDE EDGES OF
FOOTING TRENCH
DIRECTLY BENEATH
THE LINES

TOPSOIL
REMOVED

WIDTH
OF
SLAB

FOOTING
TRENCH

2'

1'

After batter boards have been installed, and the slab perimeter determined, the footing trench is dug with its outside edges directly beneath the lines that demark the edges of the slab.

Install all the services, all the pipes — water, sewer, gas — and electrical lines that have to come up through your slab and into the building, at this time.

Constructing the form. Since the slab is to be constructed on firm, dry ground, you need no more than a 2x4 on edge to provide the desired 4-inch thick concrete. However, if you want the top of your slab to be higher — remember this will be the floor of your building — construct a higher form and raise the level of the base by covering the soil with a layer of crushed stone or gravel, leaving 4 inches of space between the top of the stone and the top of the form. To keep the gravel from moving sideways into the trench place some bricks or stones (not too high) around the gravel, or slop some wet concrete around it. You can also just slope the gravel towards the trench. There is no harm in having the concrete thicker than necessary at this point; it just costs more.

The form consists simply of a large rectangular box with its inside dimensions equal to the desired slab size. In our example, these dimensions will be 30 by 40 feet. Anchoring the form may be a slight problem, since the 30-by-40-foot box may slip into the trench, which, because of the nature of soil, has to be a little larger near its

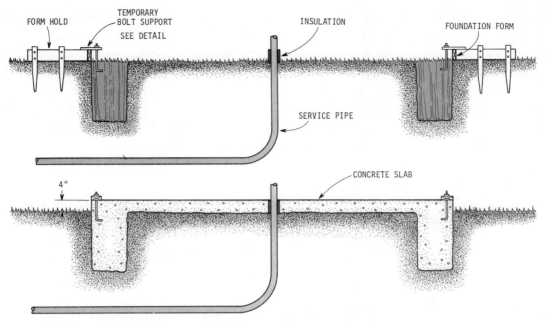

Constructing a 4-inch slab foundation directly on the ground. First, as shown in the top drawing, the topsoil is removed and the trench dug. Then service pipes are installed and insulated. Next the form is constructed and placed in position. After the trench has been filled with concrete (bottom), the balance of the form is poured, screeded, tamped, and finished just as if it were a walk or drive. Cross-sectional view shows finished slab with form removed.

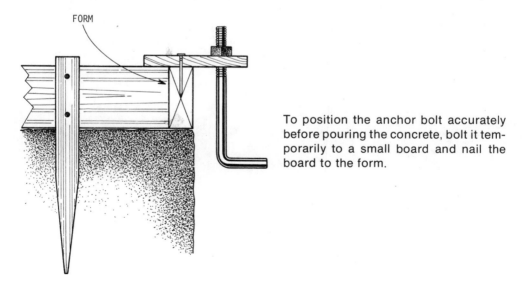

To position the anchor bolt accurately before pouring the concrete, bolt it temporarily to a small board and nail the board to the form.

top. The solution is to nail some short 2x4s along the sides of the form in such a way that their ends project and can then be anchored with stakes. Before doing this, check to make certain the form is properly in position, perfectly square and level to within an inch along its length.

Since the slab you are going to pour is fairly long, several guides should be installed across its width so that it can be screeded with a reasonably short screed. The longer the screed the more difficult it is to work with. Three cross guides should be about right; they will cut the slab into four 10-foot-wide sections.

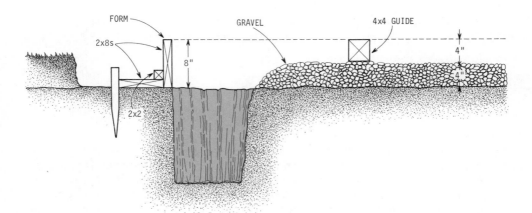

When constructing an 8-inch or thicker slab foundation, use 8-inch lumber for the form. To save on concrete, 4-inches of gravel is spread over the ground inside the form. A 4x4 laid on the gravel serves as a guide during screeding. It is then removed and the space filled with concrete. Cross-section of the finished slab is shown below.

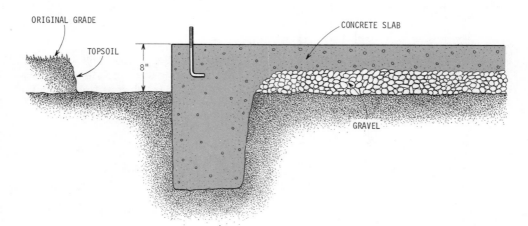

Wrap some fibrous material such as a thin sheet of styrofoam or tar paper around the pipes along the section that will be within the concrete. This will insulate the pipes from the slab so that intermovement will not place a strain on them. Later the material can be cut flush with the top of the slab.

Foundation bolts will be inserted head down in the wet concrete along the edge of the slab. They must be positioned correctly in relation to the edge of the slab, doorways, studs, and other such features. This is something that should not be done in a hurry. Take your time and carefully mark the exact position of each bolt on the form top. Install temporary bolt supports, as shown in the drawing, or press the bolt head into the partially set concrete.

Pouring. Use a 1 : 2¼ : 3 mix and a moderate quantity of water, just as you would if you were pouring a driveway or a sidewalk. If the day is hot and dry, wet the earth down, but do not soak it; soaking may soften it.

Fill the trench first, slowly and carefully. Try not to push the concrete along the trench because that may bring soil down. When the trench has been filled all around, start pouring the slab, section by section, screeding one section before you go on to the next one. You will find it easier if you pour alternative sections. Tamp right behind your screed, for a large slab like this will set up rapidly when poured directly on the ground in hot, dry country.

Finish first with a bull float and then with a hand float and metal trowel. You'll find sprinkling a little water under the trowel will help. Also, this is where you can use a power trowel to advantage. While one person is troweling, have someone else insert the foundation bolts. As soon as the concrete has set up sufficiently to hold the bolts firmly, use a trowel to scrape away whatever concrete has been pushed up around the bolts. Then, once the slab has been cured as previously suggested, you are ready to start on the building itself.

WET, WARM AREA CONSTRUCTION

Proceed exactly as before, but do not lay the slab on the ground. Instead, make a 12-inch-high form and place its inner surface under the lines you have strung from the batter boards.

Next, construct a second form within the first one, 12 inches high also, but 3 feet smaller in both directions than the outside form. In other words, if the outer form is 30 by 40 feet on the inside, the second form should measure 27 by 37 feet on the outside. The inner form should be centered inside the outer one. Make certain the outer form is perfectly level and square. The inner form need only be approximately level and square.

Once you have installed the service pipes and wires, if there are any, fill the inner form with gravel or crushed stone to within 4 inches of the top of the outer form. Tamp the stones level and smooth. Cover the stone base with a 6-mil-thick plastic sheet, running the sheet all the way from one inner edge of the form to the other. Let the sheets overlap each other by a foot or two where necessary. Mark the position of the anchor bolts on the outside form. Position three 4x4s as guides on top of the plastic water barrier. The top of the guides must be flush with the top of the outer form.

Use a 1 : 2¼ : 3 mix and a moderate quantity of water. Fill the trench and the space between the forms until the concrete is approximately 4 inches from the top of the outer form. Poke a stick into the concrete every few feet to make certain there are no air pockets. Remove the inner form after the concrete has set up for a few hours. Now pour the slab out to the edges of the outer form. Screed, tamp, float with

POURING A SLAB ON WET GROUND

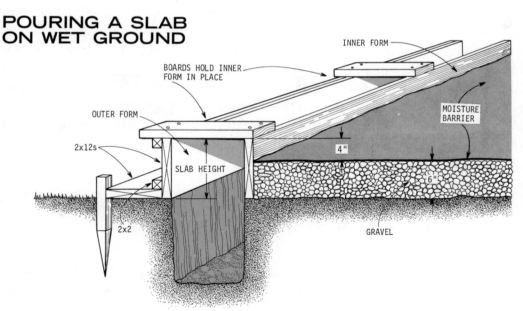

1 Construct and install two forms of 12-inch lumber, one within the other. Use short boards to hold the inner form in place. Fill the inner form with gravel to within 4 inches of the top edge. Cover the gravel with a moisture barrier.

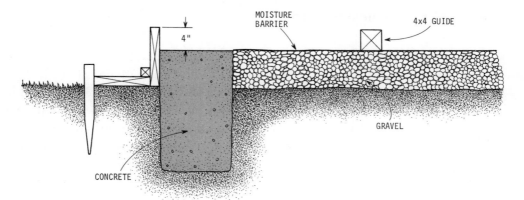

2 Pour the trench with concrete 4 inches short of full. Let the concrete set up, then pull the inner form, leaving the outer one in place. Then place a number of 4x4 guides on top of the moisture barrier.

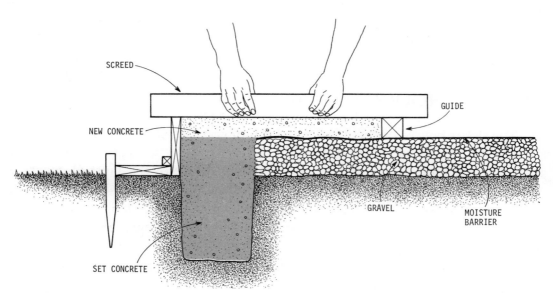

3 Pour and screed the balance of the form, removing the guides as you work.

a bull float and finish with a steel trowel. Don't forget to insert the bolts and clean up around them afterwards or to fill in the channels left by the guides after their removal. Cure as previously suggested and you are done.

FROST AREA CONSTRUCTION

Design and procedure are almost identical to those already described. The differences are that the perimeter footing—the thickened edge of the slab—must go down below the frost line and the slab itself must be insulated. The insulation suggested is suitable for moderate winters where the temperature doesn't hit zero very often and averages around 20 degrees F through the cold months. In colder areas it is advisable to secure local engineering help—the building department will or should help—before committing yourself to a slab foundation.

Laying out and excavating. Start as before. Skin the topsoil off to reach firm subsoil. Install the batter boards and lines. Dig the perimeter trench, which must go down below the frost line. Make the bottom 2 feet wide to permit you to pour a 1-foot-thick foundation wall and still have room for the form and the insulation. As previously stated, the 1-foot-wide foundation wall is sufficient for a one-story frame building on soft clay. For heavier loads and/or stronger soils see Chapters 16 and 17 or consult your building department.

An alternative to pouring a foundation consists of pouring a footing and "laying up" a concrete-block foundation wall. If you choose this method, simply fill the bottom of the 2-foot-wide trench with a very wet mix to a height of 10 inches. Use a 1 : 3 : 5 or richer mix and let it find its own level. Then lay the block on top.

The slab is poured on top of a 6-inch layer of gravel or crushed stone, topped by a 2-inch layer of styrofoam insulation, in turn covered by a layer of polyethylene sheet. The concrete is 4 inches thick. Thus the surface of the finished concrete, which becomes the surface of the first floor of the building, will be 6+2+4 = 12 inches above the bare subsoil. In this design the surface of the slab is flush with the top of the foundation wall. Thus the top edge of the footing must extend 12 inches above the surface of the exposed subsoil.

With the topsoil removed and the batter boards and lines in place, dig the trench and install the service pipes.

Constructing the form. For convenience, we will assume that the entire form is going to be constructed and poured at one time. This is the best method, but if it is too difficult for yourself and your limited crew, or if you cannot reuse the form material in your construction, you can make and pour the form in sections. When you do this, the stop boards closing the ends of the forms should have three holes in them through which you can poke short reinforcing bars. When you move the form along and pour the next section the steel bars lock one section to the following one.

POURING AN
INSULATED SLAB

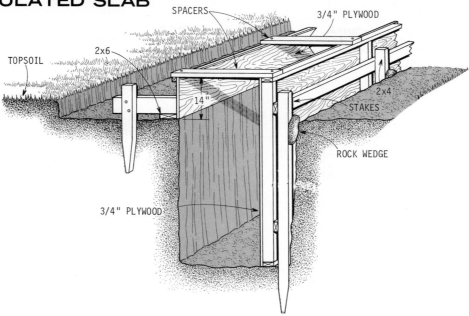

1 Dig a trench about 2 feet wide, several inches below the frost line. If the soil is firm, let one side of the trench serve as a part of the form. Construct and install an outer form 14 inches high. Construct and install an inner form of plywood on a 2x4 frame, braced as shown or as is most convenient. Pour concrete to the top of the form, screed level, and install anchor bolts.

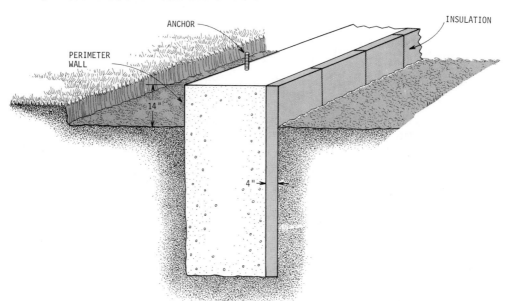

2 Remove the form and install 4-inch-thick batts of rigid insulation on the inner side of the concrete foundation, the tops of the batts flush with the top of the foundation. Place soil against the batts to hold them in place (*continued*).

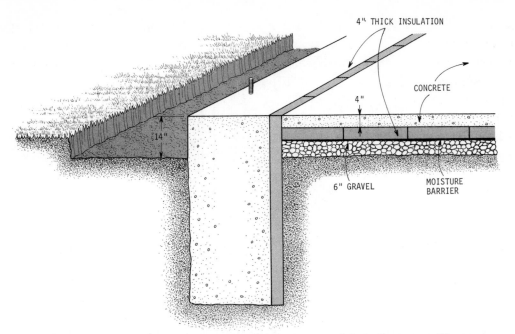

3 Cover the soil with 6 inches of gravel. Tamp the gravel smooth, cover with a moisture barrier, and lay the insulation. Then pour a 4-inch-thick layer of concrete on top of the insulation, bringing the surface of the concrete flush with the top of the foundation. Screed and finish as usual.

Construct the form with plywood or 1-inch roofers (the cheapest boards) as suggested in the next chapter. Make the inside dimension 1 foot to produce a 1-foot thick wall. Make the top edge of the form level with the desired slab surface. In our example this will be 14 inches above the exposed subsoil. Make the form level and square and position it so that the outside edges of the poured foundation will be flush with the overhead batter-board lines. Thus, if the house is to be 30 by 40 feet, the poured foundation or block wall should be *exactly* 30 by 40 feet.

Pour and screed the foundation wall and install whatever foundation bolts are required. When the concrete has hardened sufficiently, remove the form and place 4-inch-thick batts of rigid insulation against the inside surface of the foundation wall. Make the tops of the batts flush with the top of the poured wall. Do not permit any of the batts to overlap, but cut as necessary to form a continuous 4-inch layer of insulation around the inside of the foundation. Place soil behind the batts to hold them in place. Then lay down a 6-inch layer of crushed stone or gravel, tamp its surface smooth and cover with a moisture barrier that runs from one wall to the other, without breaks. Overlap the plastic sheeting or roofing paper, and run it all the way up to the insulating batts. Cover the moisture barrier with a layer of 4-inch-thick rigid insulation. Next, position 4x4s on top of the insulation and pour a 4-inch layer of concrete on top, flush with the top of the foundation. Screed and finish as usual, but hold off as long as you can before walking on the fresh concrete, and when you walk, walk gently.

19

POURED STRUCTURAL AND FOUNDATION WALLS

Poured concrete borders, which are simply very low walls, are discussed in Chapter 10. This chapter covers higher walls, which, while they too are constructed by pouring concrete into a form, require much stronger forms. This makes them an entirely different undertaking and therefore one that requires a separate chapter.

Poured concrete walls are used as alternatives to concrete block, stone, and brick for foundations for all types of structures.

Poured concrete walls are many times stronger than the alternative building materials mentioned. However, building codes will not permit you to build proportionally thinner load-bearing walls. For example, you cannot construct a 4-inch-thick foundation wall of poured concrete in place of an 8-inch concrete-block wall. It may be permissible to use an 8-inch-thick wall in place of 10-inch or even 12-inch block, but this is something that varies from one building department to another. In any event, if you want or believe you require a stronger foundation wall, the poured wall is the answer.

As for cost, there is no doubt that the concrete itself costs less than the block it replaces, but it does involve other costs that you do not have with block. The forms constitute the main expense. However, you can rent forms in some areas, and if you make your own, the wood used for them is not wasted as it can be used for flooring and roofing. The oil with which the wood forms are coated does not harm the wood. Thus there is no simple rule of thumb by which you can readily determine which alternative is going to cost you less. You have to sit down and price out everything involved.

As for labor, the forms require far less man-hours of work when you have sufficient help and when you can rent aluminum forms. When you do not have sufficient help, when you have to construct the forms of wood from scratch, the block wall is usually easier and more practical.

Typically, an experienced mason and a strong helper (helpers always do twice as much physical work as the mason) can erect a concrete-block foundation for a

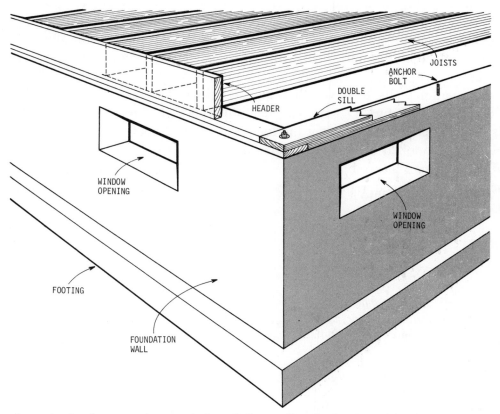

An example of a poured concrete foundation wall on its footing with sills, joists, and headers in place. Window openings are provided by installing special forms, called bucks, before pouring concrete.

small house in two or three weeks. An experienced crew, using lightweight, 3- by 8-foot form sections, can do the same basement in thirty to forty man-hours, including stripping (removing) the forms. But they erect the entire form at one time so that it is a continuous, though comparatively slow, pour. This isn't an easy task for two men, and the forms certainly represent a large investment in lumber and labor if you cannot rent them and must construct them all yourself.

FOOTINGS

Non-load-bearing walls, such as garden walls and interior walls, need not rest on footings. Their bottom surfaces offer sufficient support. If you are constructing such a wall on a concrete slab, first clean the surface of the slab, then give it a coat of a bonding agent such as Weld-Crete. If the wall is built on soil, dig down deeply enough to place at least one third of the wall's height below grade otherwise it may tip over.

Foundation walls, however, do require footings. These are constructed exactly as described in Chapter 17 with one small difference. The footing has to be keyed. This is accomplished by pouring the footing as you normally would, then pressing the length of an oiled 2x4 into the concrete and holding it there. Since the 2x4 will tend to float, nail a few boards across the form to keep it submerged. When the concrete has hardened, remove the 2x4. When you pour the wall on top of the foot-

A short, thick concrete wall can be poured directly on an existing concrete slab, but a comparatively thin, tall wall must be locked to the concrete slab with steel rods to assure adequate strength. Holes are drilled in the slab for the 1-foot rods, which are installed every 2 feet, before pouring begins.

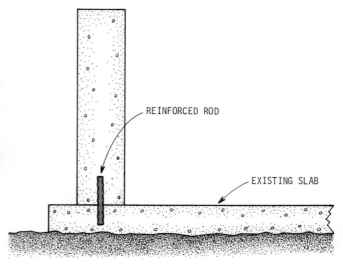

REINFORCED ROD

EXISTING SLAB

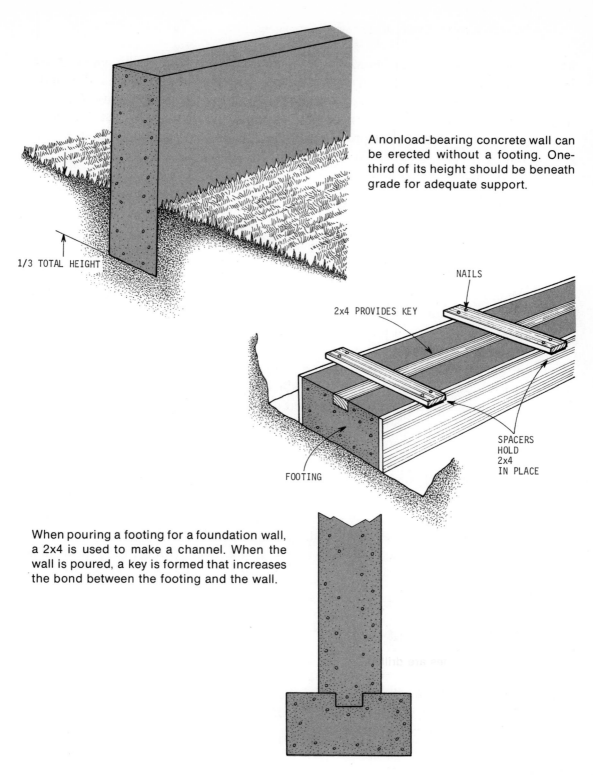

A nonload-bearing concrete wall can be erected without a footing. One-third of its height should be beneath grade for adequate support.

1/3 TOTAL HEIGHT

NAILS

2x4 PROVIDES KEY

SPACERS
HOLD
2x4
IN PLACE

FOOTING

When pouring a footing for a foundation wall, a 2x4 is used to make a channel. When the wall is poured, a key is formed that increases the bond between the footing and the wall.

ing, the bottom of the wall fits into the channel left by the 2x4. This makes for a much stronger bond between the wall and the footing than would result if the top of the footing were flat. Incidentally, if you are not going to pour the wall while the footing concrete is still green it is advisable to use a bonding agent on the surface of the footing to ensure a tight, waterproof joint.

FORM STRENGTH AND POURING RATE

The following may be a little boring, but it is necessary reading if you are going to avoid trouble.

Fresh concrete is liquid, which means it develops hydrostatic pressure that is directly related to its height. Thus, if you fill an 8-foot-high form rapidly with wet concrete the hydrostatic pressure pushing outwards against the lower sides of the form will be equal to the total weight of an 8-foot-high column of concrete, plus whatever additional weight results from the concrete dropping a certain distance downward from the ready-mix delivery chute. At a minimum the pressure will amount to 1,200 pounds per square foot, which you must admit is a lot of pressure. On the other hand, if you fill the very same 8-foot-high form very slowly and gently, say at a rate of 1 foot per hour, the pressure will never exceed 150 pounds per square foot.

The very important point made here is that pressure is directly related to pour rate. If you pour slowly and give the concrete time to set up — stiffen a little — it will carry its own weight. Since concrete sets up more rapidly in warm weather than cold, you can pour a little faster in warm weather. To bring this all into practical figures: since 1,500 is 10 times greater than 150 it is obvious that forms constructed to withstand the pressure of fast pouring must be many times stronger than forms of a similar size designed for slow pouring.

CONSTRUCTING FORMS

Forms for our present purposes are best constructed in sections. Each section consists of a frame covered with plywood sheathing and backed by 2x4 studs, in turn backed by 2x4s called walers. A pair of facing sections constitutes a form, which may be joined to other sections or be closed at each end by stop boards.

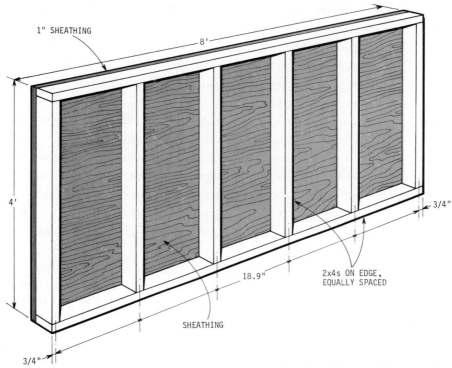

Labels on figure: 1" SHEATHING · 8' · 4' · 3/4" · 18.9" · 2x4s ON EDGE, EQUALLY SPACED · SHEATHING · 3/4"

Forms are best made in sections which can then be assembled and disassembled as required. This is much easier than building one side of a foundation-wall form at one time. The number of studs you need per form section depends on the thickness of the sheathing and the rate at which you pour. Follow the suggestions in the accompanying table to be sure your form will not burst when you fill it.

SHEATHING AND STUD-SPACING GUIDE

Sheathing thickness	Rate of pour	Stud (2x4s) spacing
$\frac{1}{2}''$	1 ft/hr	13″
$\frac{3}{4}''$	″	18″
1″	″	22″
$\frac{1}{2}''$	2 ft/hr	11″
$\frac{3}{4}''$	″	16″
1″	″	20″
$\frac{1}{2}''$	4 ft/hr	9″
$\frac{3}{4}''$	″	12″
1″	″	16″

Start by making a rectangular frame of 2x4s on edge, 4 by 8 feet overall. Cut four 2x4 studs. Make each stud 44¾ inches long. Space them equidistant from each other and from the narrow ends of the frame, then nail them to the frame. Next, sheath the frame with plywood or roofers, which are inexpensive, 1x6 or 1x8 tongue-and-groove boards. You don't need very many nails as the pressure will not

273

be on the nails. Nail the studs on edge to the frame and sheathing, using only a few nails to each one.

The number of studs already suggested and the number of walers and tie wires suggested below are based on the pressure that will be developed by filling the form with concrete at a rate of 2 feet per hour. (Note that wall thickness and length have no bearing on the pressure developed.)

To make a 4-foot-high wall one pair of sections is required. To make an 8-foot-high wall, two pairs of form sections are required because the studs run the short way across the section. One form section is nailed on top of the other through the section frame. Use double-headed nails so they can be easily pulled out afterwards. To make a form longer than a single section, nail sections end to end. To shorten a form without cutting, nail a stop board between the pair of facing sections. To go around a square corner, nail the ends of the forms to two vertical 4x4s where the forms meet at the corner.

To make a wall that is going to be longer than the section or sections you are going to fill with concrete at one time, nail a 2x4 on edge, vertically to the inside of one stop board. When you pour, the 2x4 will leave an indentation that will form a key and thus make a better, stronger bond with the concrete that you pour next. Again, if you are going to pour fresh concrete against concrete that is no longer green, it is best to coat the old concrete surface with a bonding agent.

Walers. These are braces placed horizontally behind the studs of the forms. At the 2-feet-per-hour pour rate selected, which incidentally develops a pressure of about 400 pounds per square foot, you need a pair of 2x4 walers every 36 inches of stud height. But when you place one 4-foot-high form section on top of another to make an 8-foot-high wall, you need four walers (pair of 2x4s) for the entire length of each side of the assembled forms. If you are going to pour a wall 32 feet long at one time and have erected 32 feet of forms, 8 feet high, you need eight 32-foot-long walers—four on each side, spaced evenly apart. The waler does not have to be a single piece of lumber. You can use several pieces and have them overlap by a foot or more.

Spacing and tying. Spacers and tie wires (8- or 9-gauge iron wire) are used to assemble the form sections and keep them the proper distance apart. Spacing is accomplished with lengths of 2x4 as long as the width of the wall. They are placed between the facing form sections about 2 feet apart. Thus you would need four spacers for every linear 2 feet of the 8-foot-high form.

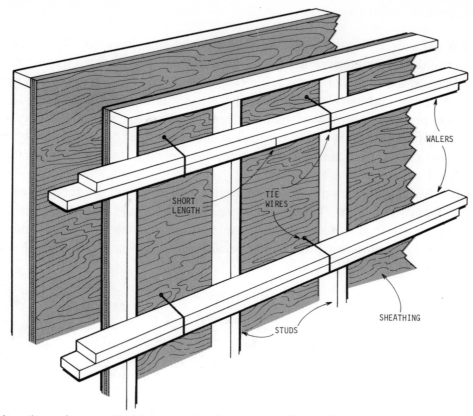

How the walers are tied to assemble the form sections. Walers can be made of short lengths of lumber butted together. Tie wires run through holes in both form sections and encircle the walers on either side.

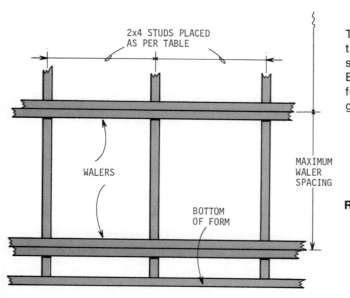

The number of walers used on each side of the assembled form sections and their spacing again depends on the rate of pour. Each waler consists of two 2x4s. Every form has at least two walers; more, if the guide calls for them.

WALER SPACING GUIDE

Rate of pour	Maximum waler spacing
1 ft/hr	46″
2 ft/hr	36″
4 ft/hr	24″

Two pairs of facing holes about ⅛ inch in diameter are drilled through the sheathing of the two facing forms. Each hole is about 5 inches from its neighbor. All the drilling is done with the forms lying conveniently flat on the ground. The holes are drilled above and below the points where the walers will be positioned.

At this point the form sections have been constructed and oiled. The spacers have been cut and the walers and wire are at hand. The next step, assembly, is best done by two people and can be carried out anywhere—at the spot where the form will be filled with concrete or nearby.

With a few nails attach the walers on both sections. Stand the sections up. Place one spacer between them. Thread the wire through a hole near the spacer, then through a facing hole, down and around a waler, back through the section and up and over the waler opposite the first. Twist the wire ends together. Now, reach into the

Walers are braced against the hydraulic pressure exerted by the concrete only by tie wires. The distance between tie wires depends on the rate of pour and the size of the wire used. Tie wires must be used on all the walers. If necessary, vary spacing to clear studs.

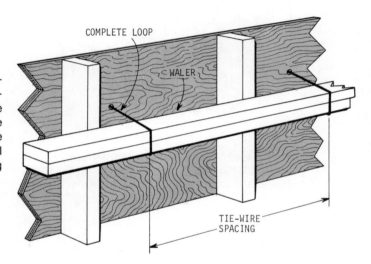

TIE-WIRE SPACING GUIDE

Rate of pour	#8 wire	#9	#10
1 ft/hr	40″	32″	28″
2 ft/hr	30″	25″	20″
4 ft/hr	24″	20″	16″

form and place a stick between the two wires. Twist the wires with the stick until they are tight. That locks the form against the spacer. Repeat until all the tie wires are in place. At the 2-feet-per-hour pour rate one tie is needed along every 30 inches of waler, with double ties at the corners. To make the spacers easy to remove as you fill the form with concrete, tie long pull wires to the spacers.

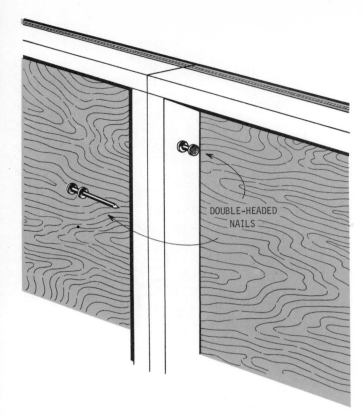

DOUBLE-HEADED
NAILS

ASSEMBLING
FORM SECTIONS

1 Form sections are first assembled by joining adjacent 2x4s with double-headed nails (for easy removal).

2 Form sections are joined at the corners by nailing end studs to 4x4s. Strips of plywood are nailed to inside 2x4s to bring 4x4 flush with sheathing. If the pour is going to be interrupted, stop boards are installed with a 2x4 attached to form a key.

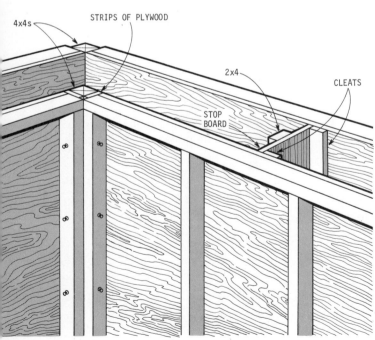

4x4s

STRIPS OF PLYWOOD

2x4

CLEATS

STOP
BOARD

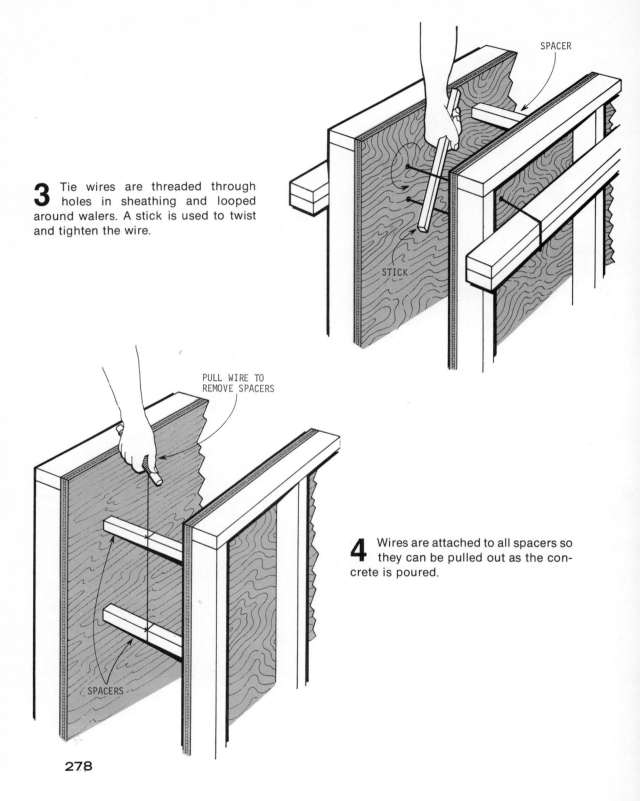

3 Tie wires are threaded through holes in sheathing and looped around walers. A stick is used to twist and tighten the wire.

SPACER

STICK

PULL WIRE TO REMOVE SPACERS

4 Wires are attached to all spacers so they can be pulled out as the concrete is poured.

SPACERS

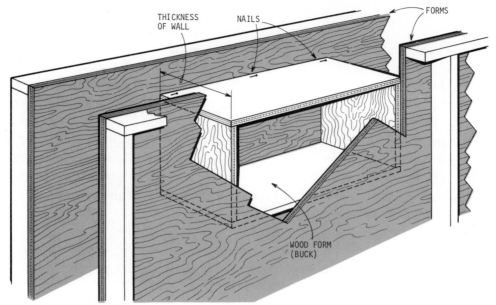

5 A wood form, called a buck, can be nailed between the forms to produce an opening for a window. When the concrete is poured, it surrounds the form, which is later removed.

6 Diagonal braces are installed and held in place with stakes. The form is now completely assembled and ready to be filled with concrete. (Stop board not shown.)

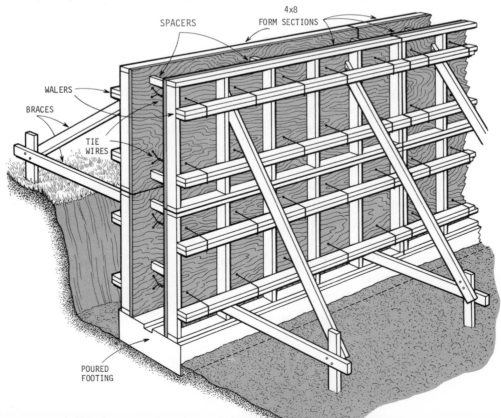

POSITIONING AND BRACING

Place the form where you want the wall to be. If you are constructing a foundation wall or a building make certain the corner is square and positioned correctly. Use a level to make certain the form is truly vertical and horizontal.

Brace the forms in a vertical position by installing 2x4s between the top waler and stakes driven into the ground. To eliminate the possibility of the form lifting up when the wet concrete hits its bottom, nail the bottom to stakes driven into the ground. Not many braces are needed because the weight of the concrete is carried by the tie wires and walers.

POURING

Windows, doorways, and other fenestrations, as well as foundation bolts, are discussed a few paragraphs further on. Pouring is discussed here, just to keep things in sequence.

When pouring a foundation or service wall that is not going to be exposed to the weather you can use the lower cost 1 : 3 : 5 mix. If the wall is to be exposed to the weather or if you want a better looking surface, use the 1 : $2^3/_4$: 3 mix. Use a moderate quantity of water — the mix should not be so loose that it finds its own level, nor so dry that you have to push it into place.

Fill the form slowly. Remember, even when you are only filling a 4-foot-high, 8-foot-long form, if you pour much faster than the 2-feet-per-hour rate the form was designed for there is a good chance you will break it apart. Excessive speed is little problem if you are mixing your own concrete, but it is when using ready-mix. Warn the plant manager of your requirement so that he allots the necessary time to the truck and so that the driver knows he will be blamed if he damages the form.

As you pour, rap the sides of the form with a shovel now and again and poke a long stick down into the concrete to help it settle and fill the voids. When the concrete reaches a spacer, pull it out. Use a rake if you haven't tied a convenient pull wire to the spacer.

You can pour a little faster in weather warmer than 70 F and should pour more slowly in cooler weather.

Give the concrete at least twelve hours to set up to sufficient hardness before you begin to remove the forms. The first step, of course, is cutting the tie wires.

Pouring concrete into a reinforced plywood form to make a footing. Although the soil is dry at this time, this is marshy ground. Footing is heavily reinforced with steel bars, and is thicker and broader than usual.

Completed, poured concrete footing and foundation. Contrary to usual practice, basement floor has also been completed. The hole in the cross wall is to accommodate pipe.

After the form sections have been removed, you can snip the tie wire ends flush to the concrete.

BOLTS, WINDOWS, AND DOORWAYS

Anchor bolts. Hang anchor bolts from a hole drilled through a short board. Nail the board to the top of the form, making certain the bolt is exactly where you need it and that the threads project sufficiently to secure the sill.

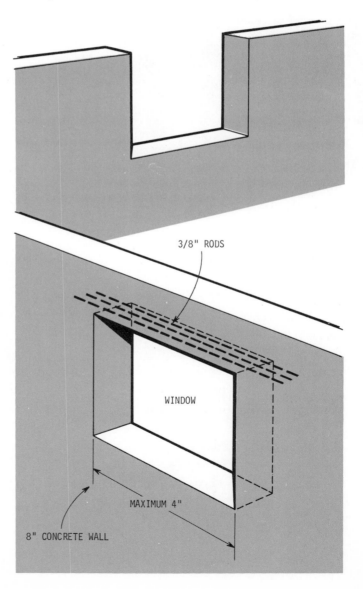

When an opening extends to the top of the poured wall, there is no need for overhead support within the wall itself (top). When the opening is within the wall but is less than 2 feet wide and runs parallel with the floor joists, no reinforcement is necessary. When the opening is wider than 2 feet or is subject to a concentrated load, it must be reinforced. Steel bars are therefore placed within the form before pouring.

3/8" RODS

WINDOW

MAXIMUM 4"

8" CONCRETE WALL

Close-up of window opening in a poured concrete foundation. Note position of sill and sheathing. The exterior of the wall will be plastered with cement. Then the siding will be applied over the plywood sheathing, the bottom edges overlapping the cement plaster by an inch or so.

Windows. The procedure is simply to install within the form a box with outside dimensions that correspond to those of the window opening. Naturally the width of the box must be exactly equal to the spacing between the form sections. A few nails will hold it in place.

Doorways. Doorways are treated the same way; the box or buck, as it is sometimes called, must be exactly the same size as the opening you want.

Lintels. So far we have assumed the door or window is either so narrow that there is no need for reinforcement above or the opening extends to the top of the form. In such cases the load is carried by some structural member above the poured concrete wall. However, if the opening does not extend to the top of the form and if it is so wide that the poured concrete above it is incapable of supporting the load resting on it, the opening must be reinforced. When the reinforcement is a separate structural member, such as a length of steel or heavy timber, it is called a lintel. In the case of a poured wall, a lintel can be made by simply laying a few bars of steel in the concrete that is poured above the opening.

The bars can be positioned prior to pouring by means of wires, or they can be laid directly on the concrete when the pour has reached the correct height. In any

283

event, the number of steel bars, their diameter and length should be specified on the building plan or cleared with the building department. There is no simple rule of thumb when the lintel must carry part of the building.

SURFACE TEXTURE

As you know, poured concrete upon hardening will reflect the molding surface. This characteristic is often used to enhance the appearance of what would otherwise be a plain flat gray wall.

If you want raised knots to be formed on the surface of your wall, use plyscore for the form. It usually has a number of shallow knot holes on one side of the sheet. To simulate the appearance of wood, use tongue and groove sheathing. The concrete will show the joints and the grain. Sheathing positioned vertically is often used for garden walls; it makes for pleasing results.

A poured concrete wall can be textured after the form has been removed by splattering it with wet mortar. The splatter can be left as is, which results in a pebbled surface, or the texture can be softened by working it gently with a wood float.

FINISHING THE TOP OF THE WALL

Level the top of the wall with a screed so the pour is flush with the sides of the form. For a smoother finish, tamp the top of the pour and then use a float.

If you want the top of the wall to be curved, mix up a batch of mortar using 1 part cement to 4 parts sand. Apply the mix to the top of the wall and then use a float or a steel trowel to shape a rounded surface.

The top of the wall can also be flagged with any suitable stone. Generally this is done shortly after the forms have been removed. See Chapter 15 for details on flagging concrete.

A concrete cap can be made by enlarging the form near its stop. This is done by nailing furring strips (narrow boards) along the top edges of the form so that the form is an inch or two wider at its top than along its sides. The method used is illustrated in an accompanying drawing. The form is filled with concrete as usual; however, the

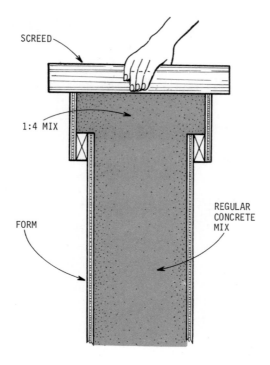

SCREED

1:4 MIX

FORM

REGULAR
CONCRETE
MIX

To cap a poured wall, the form can be enlarged as shown. A 1:4 mix is poured on top of the hardened regular mix and screeded smooth.

enlarged portion of the form, the portion that will become the cap, is filled with the 1:4 mortar mix mentioned. This produces a smoother surface and edges than would be possible with concrete.

FLOOR DRAINS

A floor drain is usually a cast-iron bowl fitted with a pierced metal cover. The bowl is set flush with the concrete. The bottom of the bowl may incorporate a trap or lead to a trap connected to a pipe that carries the collected water away.

Floor drains are installed in cellar floors that are going to be hosed down for cleaning; in cellar and basement laundry rooms, which may accidentally be inundated with water; at the low point in a driveway that serves a garage under a building and in below-grade areaways adjoining basement windows and doorways. In other words, floor drains are installed wherever rain water, wash water and floodwater cannot run off but must be led away.

Floor drains that are going to be used for driveways must be strong enough to sustain the weight of whatever cars or trucks will use the driveway. Floor drains to be used in such places as playrooms should have a flat surface so that they will not interfere with walking and will be as unobtrusive as possible. The overall height of the drain and its trap must be such that when its surface is flush with the concrete floor its outlet is still high enough, in relation to the piping, to drain properly.

In a cellar or basement, the best place for the drain is close to any corner that is out of the way of floor traffic and convenient to the house drainpipe or a pipe leading outdoors to a dry well.

In a window or cellar-door areaway, the best location is again close to a corner.

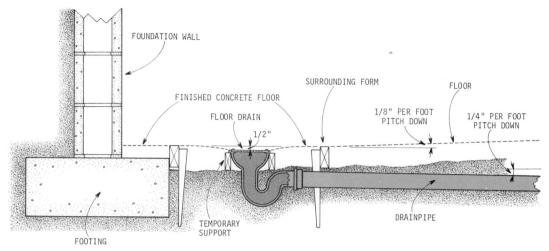

FOUNDATION WALL

FINISHED CONCRETE FLOOR

FLOOR DRAIN

SURROUNDING FORM

FLOOR

1/2"

1/8" PER FOOT PITCH DOWN

1/4" PER FOOT PITCH DOWN

TEMPORARY SUPPORT

FOOTING

DRAINPIPE

A floor drain temporarily supported and positioned within a surrounding form. Note the relation of the top of the form to the eventual finished surface of the floor.

In a driveway leading to a garage beneath a building, the best place is to one side of the lowest section of the driveway.

These positions are best because they simplify pouring and finishing the concrete floor or drive. When the drain has to be placed more than 2 feet or so away from a corner, it is usually easier if you install it in or near the center of the area.

CORNER DRAINS

Position and pitch. Temporarily assemble the drain and its pipe and position the drain for the time being about 1 to 1½ feet away from a corner and equidistant from the two adjacent walls. Use bricks or stakes to hold the drain exactly ½ inch below the proposed height of the slab at that point, which will be the lowest in the room when the slab is poured and finished. Construct a surrounding form approximately 2 feet square, using 2x4s on edge. Place two sides of the form up against the corner walls, with the drain roughly in its center. With stakes, driven into the ground inside the form, elevate the form so that its top will be perfectly flush with the finished height of the concrete at this point and ½ inch higher than the surface of the drain. (The area between the drain and the surrounding form will be hand-finished to form a small saucerlike depression.)

Use straight boards as guides to lay out the necessary pitch. The pitch in a large

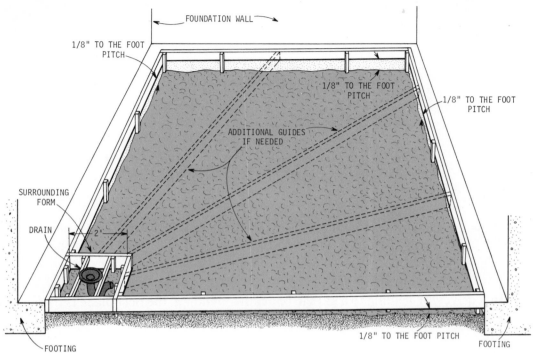

FOUNDATION WALL

1/8" TO THE FOOT PITCH

1/8" TO THE FOOT PITCH

1/8" TO THE FOOT PITCH

ADDITIONAL GUIDES IF NEEDED

SURROUNDING FORM

DRAIN

2'

FOOTING

1/8" TO THE FOOT PITCH

FOOTING

A floor drain installed in a corner of a basement floor. Guides are positioned along the walls and, if needed, radiating outward from the form, as shown by dotted lines.

room should be held to ⅛ inch to the foot, or else the finished wall height will vary too much from one end of the room to the other.

Start by positioning a guide along the whole length of one of the walls that contacts the form. The top of the guide must start flush with the top of the form and run uphill at the desired pitch. Position a second guide the same way against the other wall. Stretch a string from the end of one guide across the room to the end of the other guide. Then stretch a string diagonally across the room in the other direction, just clearing the top of the form and the first string. Mark the point where the second string reaches the wall near the corner of the room opposite the drain. You now have guides along two walls of the room, and a mark on a third wall. Next, install two more guides along the full length of the remaining walls. They should run from the ends of the existing guides to the mark on the wall.

At this point you have guides along each wall, marking the surface of the slab you are going to pour and finish. If you were to lay a giant piece of cardboard flat on these guides at this time, it would pitch down towards the drain at the desired angle.

To check the pipe and floor pitch, lay a level on the drainpipe. It should pitch down and away from the floor drain at a rate of $1/4$ inch to the foot or more. If it does not, you are going to have to raise the floor drain and everything else—form and guides. Next, check your guides against doorways and whatever piping and heating fixtures may have to be installed. For example, a furnace is always supported on a 4-inch slab on top of the concrete floor. Will the height of the proposed floor force the furnace and its piping too close to the cellar ceiling? These things must be checked out before you pour.

Assuming all the heights have proved satisfactory or have been corrected, the next step is to permanently connect the drain to its piping. Then fasten the drain in place by covering its lower end and pipe with fresh concrete. Cut the stakes flush with the surrounding form.

If you are going to use a granulated base of stone or cinders, lay it down now.

Next, position as many guides as you will need, keeping their top edges flush with the guides along the walls.

Pouring and finishing. Pour and finish as already suggested, up to and flush with the surrounding form. Wait a day or two until the concrete is fairly hard but still green. Then carefully remove the form and fill the space with fresh concrete. Using a

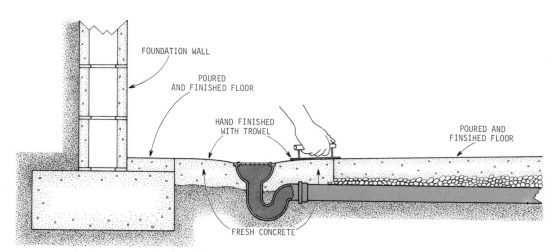

FOUNDATION WALL

POURED AND FINISHED FLOOR

HAND FINISHED WITH TROWEL

POURED AND FINSIHED FLOOR

FRESH CONCRETE

How the floor immediately around the drain is finished. First the floor itself is poured and finished. Then the surrounding form is removed and that space poured. Finally, a float is used without a guide to finish the area between the set concrete and the drain.

small float with rounded edges, shape and finish this section by hand to form a shallow bowl between the edge of the drain and the rest of the concrete.

For additional suggestions for working on concrete slabs in cellars, see the following chapter.

CENTRAL DRAINS

Position and pitch. Place the drain as close to the center of the floor or areaway as possible. If it cannot be in the very center of the space, try to position it somewhere along its central length. Temporarily connect and position the drain. Construct a surrounding form approximately 2 feet square, using 2x4s on edge, as previously described. Place the form around the drain and fasten it in place with stakes. The form should be level and ½ inch higher than the drain. Measure from the form to the farthest wall. (This and the wall opposite will be referred to as end walls.) Multiply the distance in feet by ⅛ inch and make a mark at this height on the end wall. In this way you establish a pitch of ⅛ inch to the foot between this wall and the form. Install a single guide of the necessary length, on its edge, its top running from the edge of the form to the mark on the end wall. On the end wall opposite make another mark, as high up as the first, and install a second guide from the edge of the form to this mark.

Next install two guide boards, on their sides, along one end wall, to either side of the guide already in place. These two guides should be horizontal, their tops level with that of the first guide. Two more guides should then be installed in the same way along the other end wall.

Next install guides along both side walls, perfectly horizontal and at the same height as the guides on the end walls. At this point you have guides installed all around the four sides of the room, which will establish the height of the concrete you are going to pour. A guide runs from the middle of each of the walls to the surrounding form in the center of the floor.

Check all clearances and elevations as already suggested to make certain the slab you are going to lay down will be correct in relation to doors, steps, and similar features. Check the drainpipe to be certain there is the necessary downward pitch of ¼ inch to the foot away from the fitting.

When everything is satisfactory, connect the drain permanently to its piping and lock it all in place by spreading wet concrete over the pipes and around the bottom of the drain fitting. If you want to lay down a stone base, now is the time to do so.

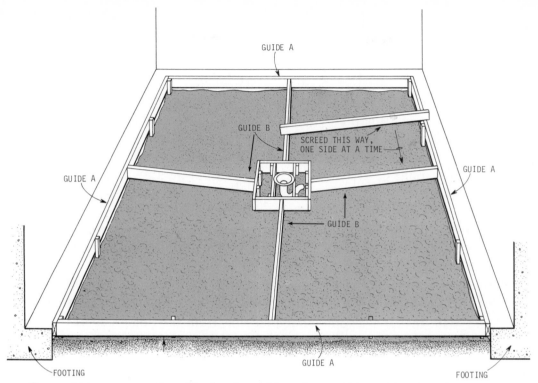

Suggested sequence of positioning the guides when working with a drain installed in the center of the floor. Note that all the guides labeled *A* are at one level. All the guides labeled *B* pitch upwards from the surrounding form to the guides along the wall.

Pouring and finishing. Starting in a corner farthest from the entrance, fill up the corner and work your way backwards. Use a screed that reaches from the guide along the side wall to the guide in the middle. Continue working and tamping and screeding as usual. Note that the end of the screed towards the center of the room goes downhill as you work towards the drain and then back upwards as you continue working away from the drain. This is the pitch that drains the floor into the drainpipe.

When you have finished half the room, let the concrete set up sufficiently and then remove the guide along the side wall and the guide along that half of each end wall. Let the center guide remain. Fill in the channels, screed and tamp, float and finish in the usual way. Wait at least a day and then do the remaining half of the floor.

The second half is a bit tricky. Use a screed that reaches from the side-wall guide to just over the center guide. Don't let the screed's end go up on the already finished concrete but keep it on the guide. When you have finished pouring, screeding, and tamping, pull the guides and surrounding form, fill in the resulting channels and finish up as previously suggested.

DRIVEWAY DRAINS

A properly designed driveway leading to a garage under a house runs downhill, then uphill before entering the building. The low point at which the drive changes elevation should be far enough away from the building to keep rain water out and to give sufficient clearance for cars at the garage entrance.

The drain should be placed at this low point, preferably, but not necessarily, near one side of the drive.

Side drains. To install a drain at the extreme side of a driveway, you must build a surrounding form. Let us assume that the left side of the drive, as you face the building, is the lower side. Here is where you would install the drain.

Start by temporarily installing the drain and pipe at the left side of the drive. Next, install a guide along the left side of the drive, starting at the form and running toward the road. Install a second guide that runs from the other side of the form to the garage.

With a level and straight board, carry the elevation of the form directly across the width of the drive and mark it on a stake. Make a second mark 3 inches higher if the drive is 10 feet wide, 6 inches higher if it is 20 feet wide. Using the higher elevation mark as your starting point, install two more guides on the right side of the drive,

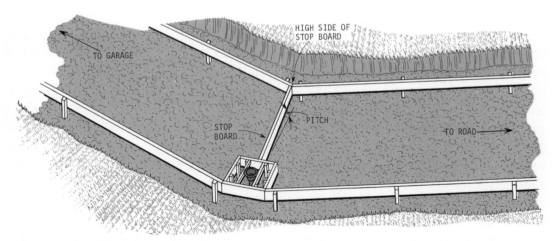

A guide arrangement used with an off-center driveway drain. Note that the stop board pitches upwards. Garage section is usually poured first; when sufficiently hard, the stop board is removed and the second slab is poured.

parallel to the first pair. The top edges of both guides should be at the same elevation as the second mark.

When the guides are in place, connect the drain permanently to its pipes and fix it in place by covering the base with wet concrete.

You now have four guides marking the perimeter of what will be two slabs of concrete pitching downhill towards one another and meeting along a line, at the left side of which is the drain. Although this line is the lowest point in the driveway, the right side of the line is considerably higher than the left, so that water falling on the drive will run into the drain.

If you are going to lay down a base, now is the time to do so. If you are going to use reinforcing wire mesh, it is laid down immediately afterwards.

The drive can be most easily poured in two sections.

Install a stop board across the drive to divide it for the pour. The board should run from the surrounding form to the opposing guide, as shown in the accompanying drawing.

Mix your concrete a little on the dry side, because you are going to pour on a hill. Then pour, tamp, and screed away until you have worked yourself backwards up into the garage. Finish the slab. Then place some clean, smooth boards on the curing concrete and pour, screed, and tamp the other half of the drive, after removing the stop board and stakes.

With a float hand-shape and finish the area around the drain. It doesn't have to be perfect, just smooth.

Central drains. Like central floor drains, these are a nuisance, and should be avoided whenever possible. But they are not particularly difficult to install.

Start by laying out the drive in three sections, after you have positioned your drain and checked it out as already described. Section one runs from the street to near the low point in the drive. Section two starts about 2 feet from the low end of section one and runs uphill to the garage. Section three consists of the space between the two low ends of the first two slabs and is divided into four small slabs. Whereas the big slabs of concrete are pitched with the length of the drive, the small slabs are pitched downward towards the center of the drive, where the drain is.

Install the side guides and the stops for the two large slabs first. Position everything so that the center of the low-end stops is about 2 inches higher than the drain on a 10-foot-wide drive and about 4 inches higher when the drive is 20 feet wide or more.

Pour, screed, and finish the two large slabs, starting with the slab closest to the garage, just as you would in the case of an ordinary drive, but remember to use

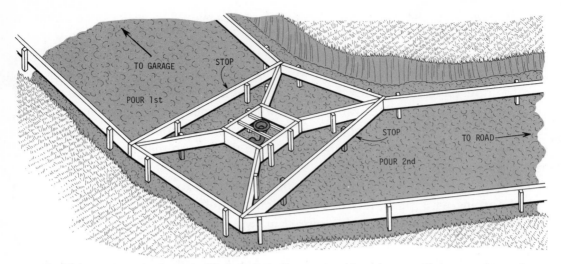

A guide arrangement used with a drain in the center of a driveway. Note sequence of pouring; the center forms are poured last.

plenty of stakes behind the low-end stops. There will be heavy pressure here from the concrete sliding down the hill.

While you are doing this, slosh a little concrete around the drain and its pipes to hold them in place.

When the slabs have hardened sufficiently for you to walk safely on them, remove the bottom stops and stakes. Install four straight boards on their sides as guides. Position each so that it reaches from the drain to a corner of the unpaved area. Make the end of the guide nearest the drain flush with the drain, and the other end flush with the surface of the concrete slab at that point. Use stakes to hold the guides in place.

Now pour and screed using a short screed and a small float. As soon as the concrete has set up sufficiently, pull the guides and fill the resulting spaces with concrete. Screed and float smooth. Then fill the inside corners between the four slabs with a little concrete and form smooth curves here with a float.

CONCRETE CELLAR FLOORS

While it is true that a concrete slab is a concrete slab anywhere you pour it, cellar slabs are sufficiently different from above-ground slabs to warrant a chapter of their own.

Cellar (and basement) slabs are laid down and finished within the confines of foundation walls and footings. This necessitates a technique different from that used with nonconfined, above-ground flat work.

Since cellars and basements are always excavated, their floors are insulated by the earth beneath and around them and by the enclosed air above them. The temperature of the earth under a cellar floor is always higher than that of the surface earth. For this reason, in addition to occupancy and heating, frost is far less likely to damage a cellar slab than a surface slab of concrete.

Since a cellar is excavated, the soil exposed is usually firm subsoil and therefore is unlikely to compact and subside as surface soil often does.

Thus neither frost nor soft soil is ever a problem when constructing a cellar or basement floor. Unfortunately, however, the very act of lowering the slab into the ground usually introduces a worse problem: water. Obviously, the deeper you go the greater the chance of encountering water. This is why many professional builders keep their cellars shallow or opt for slab foundations in areas with a high water table.

DRY EXCAVATIONS

A dry excavation is one in which no water appears either upon digging or weeks later, and from which rain water quickly drains. Assuming such is the case, you can lay your floor directly on the earth.

Wait until all the rough plumbing has been installed and the roof is up. Remove all the pipe ends, cartons, debris, and especially all the pieces of wood and wood chips. Wood embedded in concrete can absorb water and expand with sufficient force to crack the concrete that surrounds it. Remove all the loose stones and bits of mortar from the tops and sides of the footings. Remove all loose soil or tamp it firm.

Normally, a cellar or basement floor is 4 inches thick. It is best to pour the floor so that 2 or more inches of the concrete covers the top edges of the footings. This makes for a neat appearance and a better bond between the floor and the wall. However, where appearance is not important and you have no fear of water, the footings need not be covered.

If the level of the soil is 2 inches below the top of the footings and you want your concrete's surface 2 inches above the footing, there is no need to do anything. If the

Finishing a newly poured concrete cellar floor with a float and steel trowel. The form in the foreground houses the drain.

soil is lower than this or you want your concrete slab higher, spread a layer of crushed stone or gravel of the proper height over the ground. On the other hand, if the soil is flush with the footing's top or higher, you will have to remove some soil to lower the floor. Bear in mind that raising the level of the concrete floor lowers the cellar ceiling and vice versa.

All this done, you are ready to lay down the guides and forms. But since this portion of the job is identical for all cellars and basements, it is discussed after we have covered moderately wet and very wet cellar and basement excavations.

MODERATELY WET EXCAVATIONS

When the excavation retains no more than a few inches of water after a heavy rain, or when it is on the side of a hill so that collected water can be easily drained off down the hill into a storm sewer or out into a field, the following measures are advised.

Clean the cellar as already suggested. Use a stiff brush and water and carefully remove all the mud and dust that may have collected on the sides and tops of the footings and the bottom 6 inches of the foundation walls. It is important that the floor slab make a good bond here.

Remove all the loose soil from the cellar floor. Measure down from the top of the footings to the surface of the soil. If the distance is less than 4 inches, remove sufficient soil to lower the *entire* earth floor so that its surface is about 4 inches (give or take an inch) below the top edge of the footings.

Cover the earth floor with sufficient crushed stone or gravel to make a layer flush with the top of the footings. Level and smooth the stones with a metal tamper. You want to blunt all the stone points that may be sticking up so that they cannot pierce the polyethylene sheets with which you are now going to cover the entire floor. Spread the sheets—which must be 6-mil or thicker—evenly and smoothly from one foundation wall to the other. The ends and edges of the sheets must lie on top of the footing. Let the sheets overlap one another by 2 feet or more where necessary. Wear overshoes when working with the sheets so that when you step on them, there will be less chance of making holes. Some masons spread an inch or more of sand over the stone base to further reduce the chance of a sharp stone piercing the plastic sheet.

From here on the procedure is the same as for dry and very wet cellars.

VERY WET EXCAVATIONS

When the excavation is almost always filled with water; when the hole is at the bottom of a valley; when there is no way of getting rid of cellar water except by continuous pumping, seek professional advice. There are too many possible variables to attempt to cover them all here.

INSTALLING THE GUIDES AND FORMS

At this point the cellar has been cleaned and the loose earth tamped or removed. A stone base, if used, has been spread and leveled, and plastic sheeting has been positioned, assuming it is to be used.

You must now decide how and from what point you will supply the concrete to be used. If you are going to mix your own, determine where the mixing pan or machine will be placed, and if you are going to work with ready-mix, where the chute will go. This is very important because you have to work backwards. Whereas on an above-ground slab you and your crew can remain at the sides, in a cellar you start at a closed end and work backwards out the door, just as you would when painting a floor.

Since you will be inside the guides all the time you work, and since you may be removing them as you progress, the guides should be positioned in line with the source of concrete so that you do not have to step over them and can screed towards the exit.

When no plastic sheet is used, make your guides from 2x4s nailed to short pegs driven into the ground and cut flush with the tops of the guides. When using a plastic barrier, use 4x4s simply laid in place. Position the guides roughly parallel with the side walls and each other, 2 feet or so from the side walls and no more than 10 feet from each other. If the distance is greater, use more guides.

Using a straightedged board and a level or some string and a line level, make certain the upper edges of all the guide boards are as high in relation to the footings as you want them, and that they are all level or pitching in the same direction towards a doorway or a floor drain, as the case may be. See Chapter 20 for data on floor drains.

You can, if you wish, use just a few guides and move them back as you go. This, however, is a considerable nuisance if you are planning to pour the entire floor at one

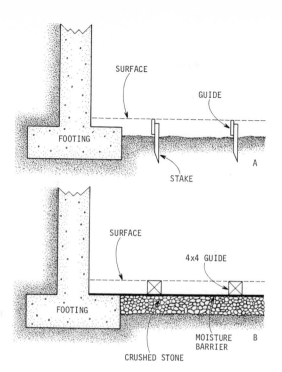

SURFACE

GUIDE

FOOTING

STAKE

A

When no moisture barrier is needed, the guides can be supported on stakes (top). When a barrier of plastic sheeting is used, lay 4x4s on top to serve as guides (bottom).

SURFACE

4x4 GUIDE

FOOTING

MOISTURE BARRIER

B

CRUSHED STONE

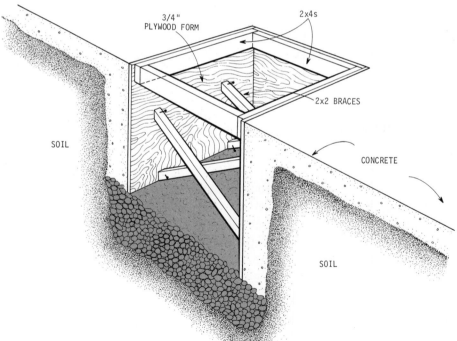

3/4" PLYWOOD FORM

2x4s

2x2 BRACES

SOIL

CONCRETE

SOIL

This form is used to produce a concrete-walled well for a sump pump. The bottom is covered with a layer of small stones to permit water to seep in.

clip. It is much easier to use several guides and position them all the way across the floor. The ends of the guide boards do not have to butt; they can overlap one another.

If there is a below-grade cleanout plug that must not be covered, construct a rectangular surrounding form of 2x4s and install it over the opening with stakes. If there is need for a sump-pump well, construct a form of plywood, as shown on the previous page, and brace it from the inside.

POURING, SCREEDING, AND TAMPING

At this point, the guides and forms are in place. If there is no stone base, the guide edges may rest on the undisturbed earth, held in place by pegs. If there is a stone base the guides rest on the crushed stone, again held in place by pegs driven through the stone and into the earth. If there is a plastic water barrier, the 4x4s just lie on it, level or pitched as required with the aid of small stones. All the forms are also in place, their top edges in line with the guides.

At a minimum you should have three strong hands on the job if you are going to work with ready-mix; four are even better. Each man and woman should have boots, a shovel, a rake, and if possible, one commercial Jitterbug (tamper).

The mix. Use sufficient water to enable the concrete to almost find its own level — there is no need for strength here. The mix should not slump flat when you dump it into the spaces between the guides, but neither should you be able to pile it more than 2 or 3 inches high. For a dry cellar floor you can use a $1 : 2\frac{1}{4} : 3$ mix with no stones larger than $\frac{1}{2}$ inch for easy positioning and working. For a wet cellar use a $1 : 2\frac{3}{4} : 3$ mix, which has more sand, because it produces a denser slab that is less permeable to water. If you wish, you can add some Anti-Hydro or a similar compound to the mix to make it even more waterproof. However, these compounds tend to speed setting time. Use the proportions suggested on the can or have the ready-mix company add the chemical.

Before pouring, wet the top 6 inches of foundation wall and the top of the footings to improve the bond between the new concrete and the old. Use water for a dry cellar and Weld-Crete or Permaweld-Z, which are bonding agents, for a wet cellar.

Begin pouring at the far end of the room. With shovels and rakes gently push the concrete to the end of the guides and up against the foundation wall. Don't push so vigorously that the concrete slops up the wall. It will remain in place there unless you scrape it off before it hardens.

When you have filled the spaces between the guides and the end wall for a distance of a foot or two, reach over and screed the concrete towards yourself using a straightedged board. There is no other practical way to do it. This is why it is difficult for an individual to do concrete floors by him- or herself.

When you have screeded a distance of no more than 2 feet, reach over with your Jitterbug or rake and gently tamp the concrete you have screeded. Then resume pouring, screeding, and tamping your way across the floor. Make certain you leave no low spots or untamped areas. Quit when you are out the door. Now, if you are mixing your own concrete, there is no problem. If you are working with ready-mix, don't let the truck go until the driver has dropped half a yard or so of extra concrete in a mixing box or even on the ground. This is important, for you are going to need this extra concrete.

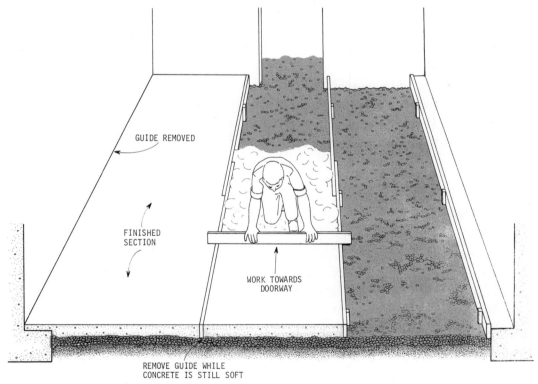

Always screed from the closed end of the room towards the door. This way you will not have to walk across fresh concrete.

Wait until the concrete has begun to set. Then walk across your screeded and tamped surface and remove the first set of guides. The forms remain in place. Be certain to remove the pegs, if you have used them. Fill the channels left by the guides with fresh concrete. Screed level with a short board and finish by tamping lightly.

Next pull up the second set of guide boards. Fill the channels with fresh concrete, screed, and tamp. If in walking you have indented the concrete, fill the indentations, screed, and tamp some more. The concrete must be smooth and unmarked by feet when you remove all the guides and prepare to finish.

Alternative methods. Instead of reaching over the screed every now and again and tamping shortly after you screed, tamping can be put off until you are at the midpoint of the floor or even later. This is a little easier and faster, but the timing is tricky. You can't wait too long before going back to tamp, or the concrete may be too stiff. And you cannot go back too quickly, or the concrete may be too soft and you will dig much of it up. Generally the best moment can be determined by a little experimentation. Concrete tends to settle downward, the bottom of the slab stiffening before the top. Therefore you will not sink very deeply into the concrete after it has been in position awhile. It should be remembered that the absence of sun and wind slows setting.

One more approach. Some masons do not tamp cellar floors at all. They depend on the high sand and water content of the mix to encourage the stones to sink below the surface. And, of course, if you aren't particularly concerned with the appearance of the finished floor, there is no need to tamp at all whatever the mix or its water content.

FINISHING

Although the details of finishing are covered in Chapter 14, there is an important point to be considered before going on. An experienced crew of four or five men can completely pour, screed, and tamp the average 1,000- to 1,500-square-foot cellar floor before the section of the floor that was poured first is ready for finishing.

Whether or not an inexperienced crew can do as well is a moot question. Therefore, if you don't have three or four men or women to help you, and this is your first time round, it is advisable to give serious thought to doing the job in two installments just to be on the safe side. Concrete is heavy and hard to move, and even a comparatively small cellar floor can be quite a job for beginners.

POURED CONCRETE RETAINING WALLS

Poured concrete retaining walls serve the same purpose as retaining walls constructed of dry masonry: they hold soil in place on a hillside in the face of heavy rains. The difference between stone and concrete retaining walls, besides the method of construction, is one of strength. The poured wall is considerably stronger. Depending on the design, the poured wall can be one quarter as thick as a stone wall of equal height and still do the job.

Three poured wall designs are commonly used. One is similar to the poured walls discussed in Chapter 19. It is simply a slab set on edge in the ground. Another is a poured slab wall set on a base. Since it may have a cross section that looks like the letter L or an inverted letter T, it is often called a toed wall. The third has a cross section somewhat like a truncated pyramid, so we will call it that for convenience.

The slab is most economical in material and labor for walls about 2 feet high in areas where it is necessary to dig 2 feet or more to get below the frost line. When there is no need to dig more than a few inches to reach firm soil, the toed design is usually most economical. And when you have lots of stone on hand and do not have to worry about frost, the pyramid design is most economical.

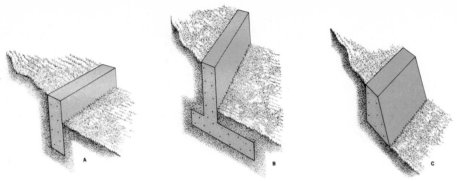

Three common designs for poured concrete retaining walls: (A) slab wall; (B) toed wall; and (C) pyramidal wall.

CONSTRUCTING SLAB RETAINING WALLS

All slab-on-edge retaining walls are made by digging a trench, erecting a form above the trench and pouring the slab. Variations in construction methods depend on the nature of the soil, the height of the wall, and the choice of form type.

The dimensions and reinforcements suggested here are for average conditions, meaning a hillside that is fairly firm and well drained. When the hill is high in clay and

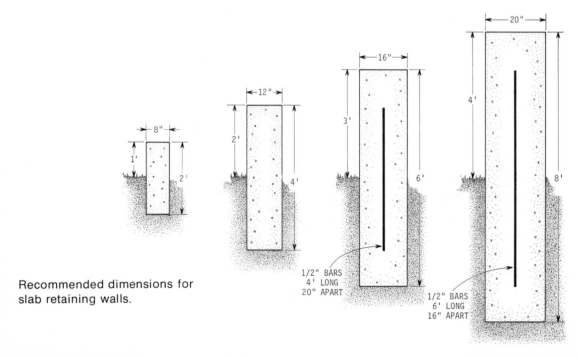

Recommended dimensions for slab retaining walls.

subject to considerable moisture, stronger walls will be required. Where the hill is not much higher than the wall itself and there is little moisture, reinforcements can be reduced somewhat.

Walls that are 2 feet high overall (1 foot above ground) should be at least 8 inches thick.

Walls that are 4 feet high overall (2 feet above ground) should be at least 12 inches thick.

Walls that are 6 feet overall (3 feet above ground) should be at least 16 inches thick and should be reinforced by ½-inch-thick steel bars, 4 feet long and spaced 20 inches apart.

Walls up to 8 feet in height overall (4 feet above ground) should be at least 20 inches thick and should be reinforced by ½-inch-thick steel bars, 6 feet long and spaced 16 inches apart.

Most municipalities require that retaining walls higher than 4 feet be designed by a professional (licensed) civil engineer. To avoid this cost, try to get the building department to suggest what they will accept. Whether or not there is an applicable building code in your area, don't go ahead on your own and build a wall higher than 4 feet. The low-wall dimensions and reinforcements suggested are for average conditions. An average hillside is not every hillside and to merely scale up the low-wall dimensions doesn't guarantee safety. If your hill is composed mainly of clay and continues upwards at a steep angle from the top of the wall, the pressure that develops when all that clay becomes soaked and fluid will be far greater than you can imagine. When a 4-foot wall collapses it is bad enough. When a higher wall gives way under a sea of mud the results are not pleasant to even contemplate. So get professional advice, one way or another, if you have to go higher than 4 feet.

Excavating. Start by cutting a level shelf into the side of the hill you plan to retain. Make the shelf as long as the projected wall plus a few feet extra for working

Steps in excavating for a slab retaining wall.

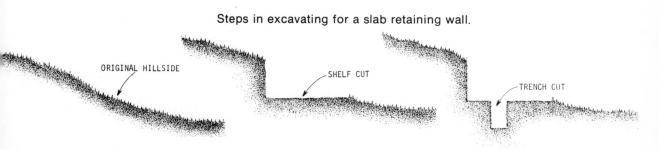

ORIGINAL HILLSIDE

SHELF CUT

TRENCH CUT

space. Make the shelf as deep as the thickness of the wall plus an additional 2 feet for working space.

Where the earth is firm and compliant, you can dig the trench no wider than the thickness of your wall and so save yourself the trouble of constructing forms that go all the way down to the bottom of the trench. When the earth keeps on spilling into the trench and when the trench has to be fairly deep, which compounds the problem, you will have to dig wide enough for the slab, the forms, and the 2x4 stakes that are needed behind the forms. A little extra width for working space will do no harm, either.

CONSTRUCTING A PLANK FORM

1 Stakes are driven into the ground alongside the trench. Planks are placed on edge between the stakes and held apart by spacers. A tie wire is looped around the stakes and tightened by twisting.

2 A second pair of planks are placed on the first pair and joined by a tie wire and spacer.

306

Constructing the form. When the form's total height is under 4 feet, the easy way to construct it is to use 2-by-12-inch planks on edge for the sides. Stakes are driven into the ground and the planks are lightly nailed, one above the other, to the stakes. To withstand the push of the wet concrete, facing pairs of stakes are tied together with loops of 8- or 9-gauge iron wire. A tie wire should be used at every stake above every plank. Spacers—2x4s cut to the same length as the thickness of the wall—are placed between facing planks near tie wires so that as you twist the ties and take up on them the stakes and planks are pressed against the spacers, locking the assembly firmly together.

3 Cleats are nailed inside the ends of the planks.

4 Stop boards are installed behind the cleats. The stop boards terminate the pour and give the end of the slab a neat finish.

307

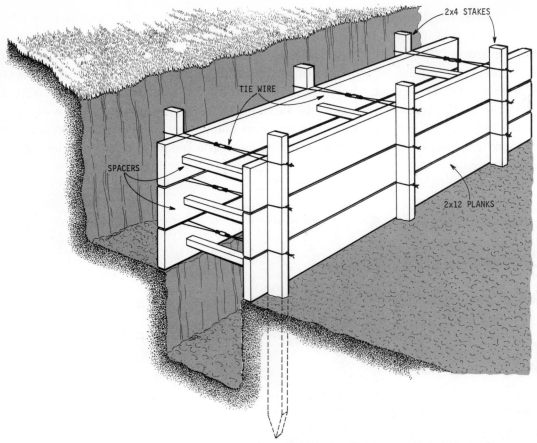

End view of a completed form constructed of planks and braced by 2x4 stakes.

When the height of the form reaches 4 feet, it is usually easier to work with form sections made of 2x4s sheathed with either 1-inch tongue-and-groove boards or plywood. The construction and bracing of these forms are discussed in Chapter 19.

Pouring. Oil the inside of the form with new or used crankcase oil. Use a 1 : 2¼ : 3 mix for the entire pour. Or, if you wish to save a little money, use a 1 : 3 : 5 mix for the below-ground portion of the slab. Pour very slowly, no more than a foot of form height per hour unless you brace the form accordingly. If concrete oozes out from a space between the bottom of the form and the ground, hold up the pouring and cover the escaping concrete with soil. This will draw water out of the concrete at that point and stiffen it up, permitting you to resume pouring without losing concrete.

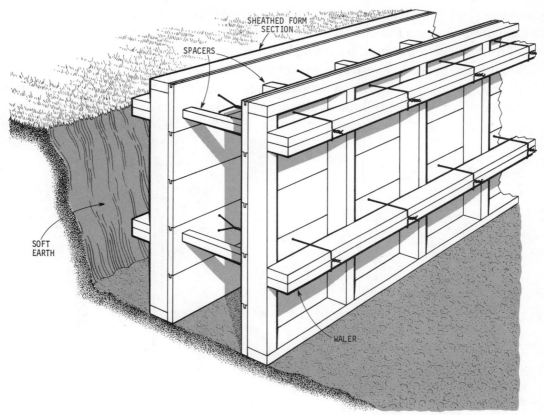

When the earth is too soft to be used as a form, the form itself must stand on the bottom of the trench.

Finish by screeding the top of the form, tamping the concrete and floating it smooth with a trowel. Give the concrete sufficient time to set up before removing the form. And don't forget to snip the tie wires first.

CONSTRUCTING A 3-FOOT-HIGH TOED WALL

When the vertical section is not going to be more than a few inches higher than 3 feet, it need not be more than 18 inches thick. The base for this wall should be about 3 feet wide and 18 inches thick. The joint between the vertical section and the base should be reinforced with ½-inch-thick steel bars, 4 feet long, bent to a right angle in the middle and spaced about 20 inches or less apart.

The wall section can be positioned near or at either edge of the concrete base. Maximum resistance to overturning is secured when the base is beneath the hillside. To do this, however, you have to remove a great deal of dirt, and the difference is not worth the effort. So position the wall section where it is most convenient for you.

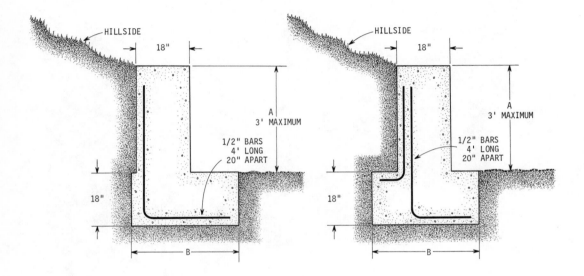

Recommended dimensions for toed walls. Note that the vertical section may be anywhere on top of the base. The height of the vertical section (A) is always equal to or smaller than the width (B) of the horizontal section. The base and vertical section should be the same thickness.

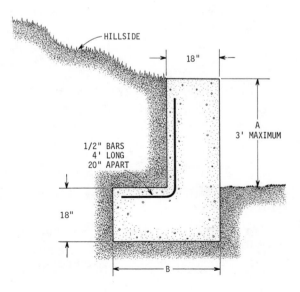

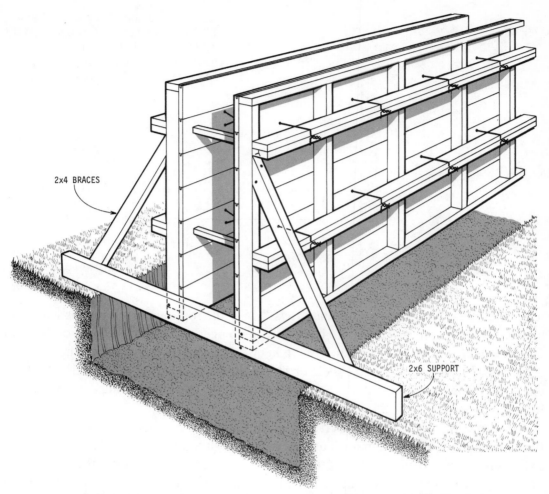

2x4 BRACES

2x6 SUPPORT

How to eliminate the need for a footing form when making a toed wall. The 2x4 supports and braces hold the form upright above the trench for the footing.

When the wall is near one edge of the base, turn the steel bars all one way. When it is near the center of the base, alternate the direction in which the bars are turned so that there is some steel in the base on both sides of the wall section.

Start by cutting the hillside back and leveling the earth alongside. Dig a trench for the footing alongside the hill. If frost is no problem, dig down just deep enough to reach firm soil, or as deep as necessary for the required footing thickness plus a few

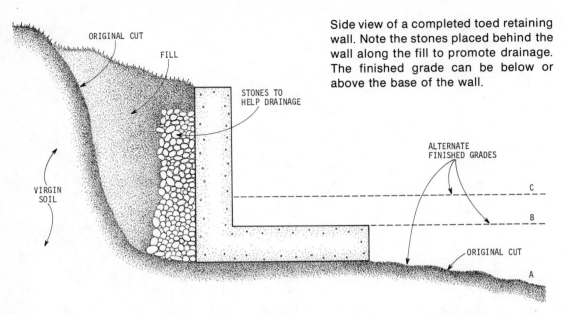

ORIGINAL CUT

FILL

VIRGIN SOIL

STONES TO HELP DRAINAGE

Side view of a completed toed retaining wall. Note the stones placed behind the wall along the fill to promote drainage. The finished grade can be below or above the base of the wall.

ALTERNATE FINISHED GRADES

C

B

ORIGINAL CUT

A

inches to allow for covering soil and grass. Make the footing trench just as wide and as long as you want the footing to be. There is no need for a form unless you have to dig very deep and the sides of the trench keep falling in. In such cases either make a form out of 6-inch planks and stakes or fill the entire hole on the assumption that the extra concrete is cheaper than the cost and trouble of constructing a form for the footing.

Prepare the form for the vertical section as already suggested. Assemble the form completely, or as much of it as you can or wish to handle at one time. Nail 2x4 extensions to the bottom sides of the form. Now, slide the form over the trench and position it where you want it. Use diagonal braces and stakes to hold the form upright and perfectly horizontal, assuming you want it horizontal.

Next, position the steel bars using tie wire to wire them into place. Oil the inside of the form and then you are ready to pour concrete.

Fill the footing section first. You can use a 1 : 3 : 5 mix here if you wish, with the stronger 1 : 2¼ : 3 mix following. Hold off additional pouring for thirty minutes or so, then add a little concrete to the vertical section. If the fresh concrete sinks into the concrete already in place, you have to wait a little longer. As soon as this stops happening, you can slowly pour the vertical section.

Toed walls 4 feet high require a base at least 4 feet wide and 20 inches thick. The vertical portion of the wall should be at least 20 inches thick and reinforced every 16 inches by ½-inch-thick steel bars, 4 feet long and bent at their centers.

CONSTRUCTING PYRAMIDAL RETAINING WALLS

Pyramidal retaining walls resist the push of the hillside by their mass alone. Therefore you can use the less expensive 1 : 3 : 5 mix for the entire wall and even add as many *clean* stones of any size and shape as you wish to reduce the cost of the concrete even further. The wall doesn't have to be strong. Where the frost is mild pyramidal retaining walls can be constructed without extending their bases into the earth. But where the frost line is 1 foot or more below the surface of the earth, you have to excavate to within at least 6 inches of the frost line for the footing to be reasonably free of the danger of frost damage.

Design. The dimensions of a pyramidal retaining wall are not critical. You can vary the suggested dimensions considerably and still be safe unless you are trying to retain a monstrous hill composed mainly of clay. In such cases, as suggested previously, secure local engineering aid.

Walls under 4 feet in height should be 30 inches wide at the base and 10 inches at the top. Walls higher than 4 feet but not over 5 feet should be 36 inches wide at the base and 10 inches at the top. If necessary, you can increase the overall height of the wall by extending the base into the earth for a foot or so. In such cases the 4-foot wall may rest on a base that is 30 inches wide but no more than 12 inches deep. The base will, of course, be poured at the same time as the rest of the wall. The 5-foot-

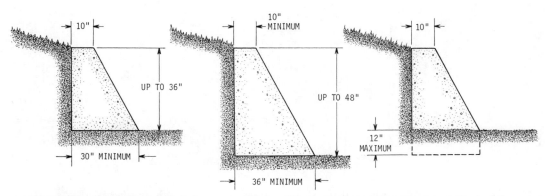

Recommended dimensions for pyramidal walls. Note that you can extend the height of the wall safely by 1 foot if that portion of the wall is below grade. Otherwise, the wall normally rests on leveled subsoil, uncovered when the side of the hill was shelved.

high wall may rest on a 36-inch-wide base that extends no more than 12 inches into the ground.

Construction steps. Start as already suggested by cutting a shelf in the hillside. Dig a trench for the base section if necessary. Construct and erect the vertical portion of the form, using heavy stakes to hold it erect. The sloping sections will have to be a sheathed form section. Hold its bottom edge in place with stakes. Tilt it towards the vertical side of the form. Use wood spacers to separate the tops of the forms the desired distance. Hold the forms against the spacers with tie wire, as already suggested.

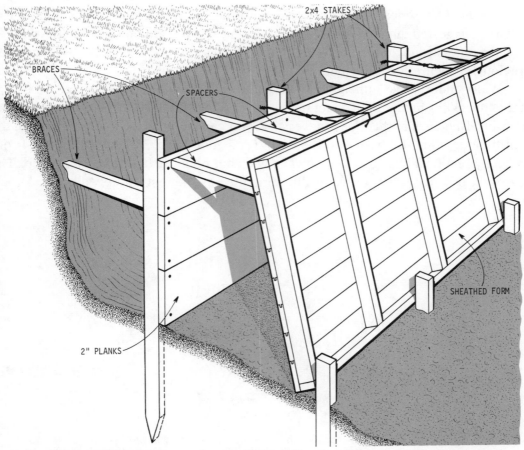

How a pyramidal wall form may be constructed. Use tie wires to hold the top edges of the form sections together, and pour slowly.

Pouring. Oil the inside of the form. Pour slowly. After you have a few inches of concrete in place, toss the rocks in, taking care to keep them away from the sides of the form—unless you want them to show. Since there is nothing more than the tie wires at the top of the form sections to hold them together, keep the pour rate down to less than half a foot per hour.

DRAINAGE

Except for retaining walls constructed indoors or out on a desert, all retaining walls higher than 2 feet must be drained. Otherwise water collects behind them, softening the soil, increasing the pressure on the wall and in the event of frost, often pushing the wall away from the hillside an inch or two every season or cracking the wall.

There are two ways to drain a poured retaining wall. One method consists of placing drainpipes behind the wall. This makes for a neat, invisible job. It is called

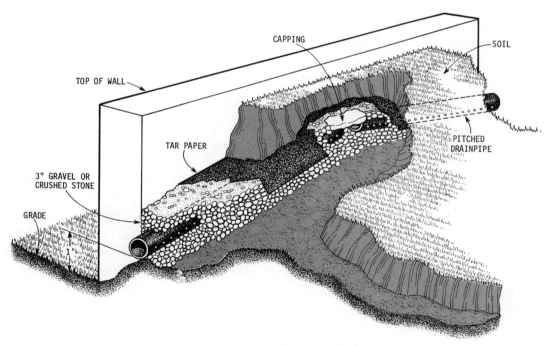

Lateral drainage behind a retaining wall.

315

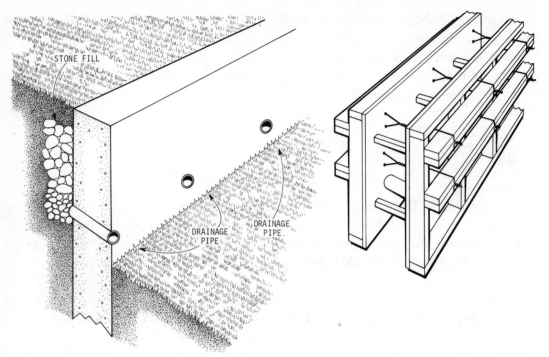

Transverse drainage in a retaining wall. Drawing at right shows how the drainpipes can be positioned within the form during the pour.

lateral drainage. The other consists of running drainpipes through the wall. The pipe ends are called weep holes and the method is called transverse.

Lateral drainage. Remove the form as soon as the concrete has set up sufficiently. Fill and lightly tamp the space behind the wall with soil to a height of about 1½ feet above grade at the approximate center of the wall and about 1 foot above grade near each end of the wall. Cover the space you have just tamped with a layer of crushed stone or gravel about 3 inches deep, 12 inches wide and just as long as the wall. Place two or more drainpipes—3-inch bituminous fiber or plastic pipe with holes drilled along its sides—on the crushed stone. The two or more lengths of pipe should pitch downward from the center of the wall and extend beyond its ends. Close the high ends of the two pipes with large stones but leave the lower ends open. Cover the pipes with a 3-inch layer of crushed stone, followed by a layer of tar paper. Now you can complete backfilling the space behind the wall.

Water seeping through the hillside will also seep through the crushed stones, enter the pipes and run out through the ends.

Transverse drainage. This is an alternate method of draining the water from behind a retaining wall. Cut a sufficient number of lengths of 3-inch plastic pipe to install one weep hole every 5 feet or so of wall length. Make each pipe length exactly equal to the thickness of the wall. When the form is in place, insert the pipe sections into it, spacing them evenly apart and about 1 foot above grade, as shown in the accompanying drawing.

PART III

WORKING WITH CONCRETE BLOCK

Not too many years ago it was possible to purchase all the lumber you needed to construct a seven-room house for under a thousand dollars. Not only was every board precut to size, the lumber company offering the knock-down home guaranteed that there wasn't a knot in the carload of wood. Those days are gone and not likely to return in our lifetime. Lumber is no longer the least expensive building material available. Concrete block is. Block is strong and not only is it fire-, vermin- and rot-proof, it is almost indestructible.

Block is no longer simply smooth and rectangular. It is now manufactured in a variety of patterns. A building constructed of block no longer looks like a factory. You can also cover the exterior of a block building with stucco, smooth or textured. And you can veneer (cover) the surface of a block wall with brick or stone.

Block's major drawback, its high thermal conductance, can be corrected with proper insulation. And block's rough surface texture can be eliminated inside the building by plastering its surface or furring it out and nailing Sheetrock over the strips.

In short, if you are thinking of constructing a strong, permanent wall or building inexpensively, give serious consideration to building it with concrete block.

TOOLS FOR WORKING WITH CONCRETE BLOCK

Line level
Pointing trowel
Finishing trowel
   Plus . . .

Brick hammer

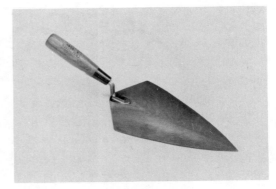

Brick trowel

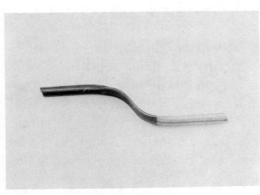

Joint tool

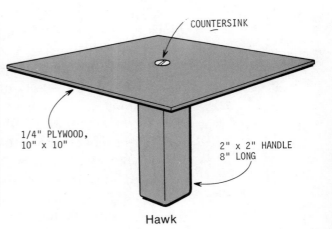

COUNTERSINK

1/4" PLYWOOD,
10" x 10"

2" x 2" HANDLE
8" LONG

Hawk

LAYING CONCRETE BLOCK

Although it takes skill and experience to lay concrete block rapidly, with perfect joints throughout, almost anyone can lay block slowly and neatly so the result is as strong as a professional job.

While an experienced mason will often lay a course or two by eye alone, especially if they are at the bottom of a foundation wall where they won't be seen, a beginner can do just as well, albeit at a considerably slower pace, using mechanical aids such as string and level. Furthermore, don't be surprised if you find yourself laying block like an expert (or thinking you can lay block like an expert) before you have even laid up half a dozen courses.

TYPES OF BLOCK

There are perhaps some thirty standard types of concrete block and innumerable ornamental and special kinds currently manufactured. However, ninety percent or more of all the block used consists of just two kinds, not counting sizes. So we shall concentrate first on these.

One of the two types of block is called a stretcher block and is easily recognizable by the indentations at both ends. This block goes in the middle of a wall. It may have two or three cores, or holes. The three-holer is a little stronger, heavier, and more costly than the two-holer. For everything any do-it-yourselfer is likely to construct, the two-holer is quite adequate. However, some building codes insist on three-hole concrete blocks. Some also will not let you build a house foundation with anything smaller than 12-inch-wide blocks.

The sides of the block are called face shells; the cross pieces in the middle of the block are called webs.

The second kind of block most frequently used is called an end or corner block. There is an indentation at one end only. Otherwise it is similar to the stretcher in all respects and can be used in place of it, should you wish to do so. Corner blocks are used at the end of a wall, where you do not want the indentations to show. If you have no corner block handy, you can use a stretcher and fill the end with mortar.

Unless you specify otherwise, most masonry supply yards will automatically send you a 50-50 mixture of corner and stretcher blocks.

A third type of block, not as frequently used, but still considered a standard type, is the solid. The most common is the 4-inch solid, followed by the 6-inch solid. They are used to cap foundation and other walls; to carry concentrated loads, such as the end of a steel girder; and to fill in spaces that would otherwise require cutting a cored block.

A fourth type, used infrequently, but worth knowing, is the "brick." This is a concrete block made in the size of a standard, common brick. It is used for such functions as filling spaces and carrying concentrated loads.

CONCRETE BLOCKS USED IN CONSTRUCTION

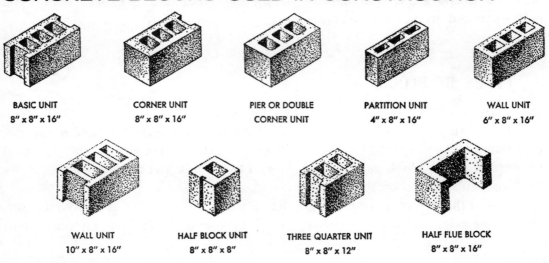

BASIC UNIT
8″ x 8″ x 16″

CORNER UNIT
8″ x 8″ x 16″

PIER OR DOUBLE
CORNER UNIT

PARTITION UNIT
4″ x 8″ x 16″

WALL UNIT
6″ x 8″ x 16″

WALL UNIT
10″ x 8″ x 16″

HALF BLOCK UNIT
8″ x 8″ x 8″

THREE QUARTER UNIT
8″ x 8″ x 12″

HALF FLUE BLOCK
8″ x 8″ x 16″

SOLID UNIT
2" x 8" x 16"

PARTITION UNIT
3", 4", 5", 6" x 8" x 16"

¼ and ¾ SASH UNIT
4", 12" x 8" x 16"

FULL SASH UNIT
8" x 8" x 16"

CORNER BULLNOSE UNIT

DOUBLE BULLNOSE, END

DOUBLE BULLNOSE, SIDE

SINGLE BULLNOSE UNIT

ALL BULLNOSE BLOCKS ARE MADE IN VARIOUS WALL WIDTHS

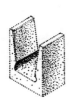 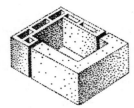 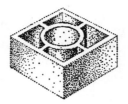 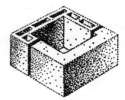

BEAM LINTEL UNITS
VARIOUS WIDTHS, HEIGHTS

VARIOUS TYPES OF CHIMNEY BLOCKS — CHECK LOCAL
SUPPLIER FOR TYPES AVAILABLE, FOR SIZES OF BLOCKS, OPEN OR SOLID

  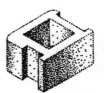

DOOR JAMB UNIT
FULL BLOCK 8" x 8" x 16"
ALSO HALF BLOCKS

SOFFIT FLOOR FILLER UNIT
8" x 21" or 24"
VARIOUS HEIGHTS

PILASTER AND HALF PILASTER
FOR ALTERNATE COURSES

DRAIN TILES, 8" LONG
4", 6" OPENINGS

COPING BLOCK
IN WALL WIDTHS

HALF-HIGH UNITS
4" x ¼", ½, ¾ LENGTHS

"TILE" UNITS, 12" LONG
8" WIDTH x 5" OR 4" HEIGHT

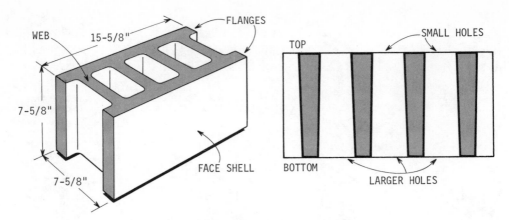

Parts and dimensions of a standard concrete block.

Now we come to the special and ornamental block. The most useful special block for house construction is the chimney block. This is a large, square block with a central opening for a flue. You simply lay one on top of another to make a chimney, so that one chimney block takes the place of four standard blocks. Other special blocks are designed for such things as framing doors and windows. They are excellent, but rarely used because they must be positioned to within a fraction of an inch, and this is something masons avoid as much as possible. It is easier to cut and patch. There are jamb, lintel, sash, double-corner, bullnose, full-cut and half-cut header blocks, just to name a few of the special blocks that are sometimes used in concrete-block construction.

The number of different types of ornamental block available is, paradoxically, both large and limited. There are lots of types manufactured, but unless you are prepared to pay for long-range hauling, your choice is more or less limited to what the local concrete-block plant produces. Ornamental blocks are discussed further in Chapter 26.

Block weight and material. Standard blocks are made of concrete using small crushed stones. A standard block measuring 16 by 8 by 8 inches weighs between 40 and 50 pounds. Figured on the cube, standard block weighs about 60 pounds per cubic foot.

Lightweight or cinder blocks, as they are often called, are made of concrete using cinders, volcanic ash, and similar porous and lightweight aggregates in place of stone. Lightweight blocks weigh about 40 pounds per cubic foot, with a block 16 by 8 by 8 inches weighing between 25 and 35 pounds.

Lightweight blocks should not be used for foundations or similar constructions because they are porous and not as strong as standard blocks. Many building codes prohibit their use for such applications. However, these blocks are excellent for non-bearing partition walls, garages, chicken coops, and similar uses.

ESTIMATING BLOCK QUANTITY

It is most important to remember that a concrete block is always $3/8$ inch smaller in dimension than its designated, or nominal, size. For example, a 4-inch block whose dimensions may be given as 4 by 8 by 16 inches is actually $3^5/8$ by $7^5/8$ by $15^5/8$ inches. This is because the block is intended to be positioned with the aid of $3/8$ inch of mortar. Thus a wall consisting of ten courses of 8-inch-high block will be 10 by 8 inches or 80 inches high because there will be nine layers of $3/8$ inch-thick mortar between the blocks, plus another $3/8$ inch under the bottom row of blocks.

The easiest way to estimate block quantity, and the one least likely to lead to error, is to work with inches. Divide the total length of the wall, in inches, by the nominal length of the block you are planning to use. When the walls includes corners, use the overall, outside dimensions. This will give you the few extra blocks you need to take care of those broken in delivery and those you break while trying to cut them.

Multiply the number of blocks you need for each course by the height of the wall, in inches, divided by the height of the block.

Figure the wall as solid, then subtract any opening, but only if it is larger than 3 by 3 feet or so. Multiply the height of the opening by its width in inches, then divide by the face area of the block. For example, an 8- by 6-foot opening works out to be 6,912 square inches. An 8- by 16-inch block has a face area of 128 inches. Dividing 6,912 by 128 (which is easy with a pocket computer) produces 54. Thus, because of the opening, you need fifty-four less blocks. However, since you will have to cut and hack to produce the opening within your wall, figure only forty-four less block, leaving ten extra blocks for contingencies.

325

MORTAR

The mortar used for all types of concrete block is a mixture of cement, sand, lime, and water. The strongest mixture consists of 1 part cement, ¼ part hydrated lime, and 3 parts sand by volume. (Incidentally, reducing the percentage of sand does not measurably increase the strength of the mortar.) This mixture is called type M by the American Society of Technical Materials (ASTM), and unless another mixture is specified by your building department, this is the one to use to be on the safe side.

You will need approximately 1 cubic foot of mortar for every ten to fifteen concrete blocks, 8 by 8 by 16 inches in size. Since particles of lime and cement fit between grains of sand, figure 1 cubic foot of mortar will result from every cubic foot of sand and other materials you mix together. Thus you can make 3 cubic feet of mortar from each bag of cement, which when added to lime and water will enable you to join thirty to fifty blocks—depending on how generously you apply the mortar.

One point before going on. In addition to the various cements discussed in Chapter 6, there is a cement called mortar cement. This is standard cement to which approximately ¼ cubic foot of lime has been added. This means that to make, for example, 3 cubic feet of type M mortar, you simply mix one bag of mortar cement and 3 cubic feet of sand.

Mixing. This is simple enough. Measure the required quantities into your pan or mixing box. Mix the ingredients dry (or the lime will stick to the hoe or shovel), then add water and mix thoroughly until the color is consistent throughout. The difficult part is correctly estimating the quantity of water to use; this is critical and will greatly affect the ease with which you can lay the block.

In order to fully appreciate the importance of water quantity it is necessary to know the whole story. As you may know, the greatest adhesion between mortar and concrete block (and brick) occurs when the block is wet. However, it is very difficult to work with wet block. The mortar remains moist and soft for a long time and the blocks tend to shift under hand as you work. To combine good adhesion and the quick stiffening of the joints that results when the block is dry, a stratagem has been developed over the years: The mortar is made wet, so that it moistens the block. The dry block draws the water out of the mortar, stiffening the mortar somewhat and also producing the desired adhesion. By using wet mortar and dry block a mason can lay up several courses without the wall slumping down and without having to wait between courses for the mortar to dry.

MORTAR MIXES BY VOLUME

ASTM Designation	Cement	Hydrated Lime	Sand	Maximum Permissible Compressive Loading in psi*	Recommended Service
M	1	¼	3	400	Good workability Very high strength Good for exposed service
S	1	½	4½	350	Excellent workability Moderate strength
N	1	1	6	300	Excellent workability Low strength
0	1	2	9	250	Excellent workability Smooth and highly adhesive Very low strength

* These figures assume that the mortar will be used with materials rated at 800 psi compressive strength or more. When used with weaker material, permissible loading is proportionately reduced. Note that standard concrete block has a compressive load rating of 800 psi or more.

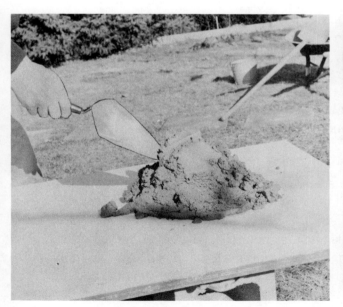

Use a piece of plywood or a mortar pan supported by a couple of blocks to hold the mortar while you work.

To get the correct quantity of water into the mortar mix, add the water slowly while mixing continuously until all the mortar in the pan or box is one color and the mixture has the consistency of stiff ice cream. The same water rule applies to mortar as to concrete: A little more is always better and safer than too little.

After you have mixed the mortar let it sit for five or ten minutes and then mix it some more. This breaks up the initial stiffening and helps you make a better estimate of the quantity of water you have added. If the mortar has stiffened considerably after a few minutes, you need more water. If it is about the same consistency as before, you have added sufficient water.

JOINT THICKNESS

Blocks are modular, meaning they are made in regular sizes. Thus, if you plan your wall correctly, there should be no need to cut any block. Or, if the wall isn't an even multiple of a standard block and joint thickness, theoretically you can still construct it without cutting block by varying the joint thickness. In practice, however, this doesn't work out unless you are very skilled or spend an inordinate amount of time just making certain each joint is the desired thickness. The far more practical way, unless the blocks are going into a front wall that is to be exposed to direct gaze, is to proceed normally, simply eyeing the joints, and correct for variations by cutting one block. This is the last block in the row, generally positioned somewhere in the middle of the row or course and called the closure block.

Therefore, don't worry about modular design; don't go to the trouble of laying out a course of blocks just to see how much of a joint you need between them. It is a waste of time.

The same holds true for height. Block is $3/8$ inch short to allow for $3/8$ inch of mortar between courses. However, joints between courses are usually $1/2$ to $5/8$ inch thick because this is easier and it doesn't matter much if a foundation or garden wall is a few inches higher than designed. When it does matter, as for example, when installing block under an existing porch, use a story pole. This is a pole marked off in even courses and placed vertically alongside the work. As you work your way up, course by course, you check your progress against the story-pole markings and so can tell if you are laying down too much or too little mortar. The marking on the story pole can be adjusted to any joint thickness necessary.

PREPARING FOR THE JOB

Start by standing the blocks on their ends alongside the wall or whatever else you propose to construct. This is done to save you the labor of carrying each block from the delivery point to the wall every time you lay a block. The mortar is placed either on a mortar board, which can be a 2½-by-2½-foot square of plywood or a commercial board that you have purchased or rented, or else on a mortar pan. To eliminate unnecessary bending, the pan or board is raised about 2½ feet above the ground using either a rented mortar-pan support or some concrete block. If you are working a long course, set up several pans along the wall so that you do not have to take more than a step or two from the pan to the wall. Keep a bottle or can of water beside the mortar pan so that you can sprinkle a little water now and then over the mortar as it dries. When you do this, work the water into the mortar with a trowel. If any mortar remains unused for two hours after initial mixing, it is best to dump it even though it may still be soft and workable.

LAYING UP THE BLOCK

With the mortar mixed and placed where it is handy, with the block standing in position alongside the job, you are ready to go to work.

Let us assume you are going to lay up block to make a wall on a concrete footing. For convenience, let us call one side of the wall its front side, the other its back side. And let us call the end of the wall at which we begin the near end and the other the far end.

Start by marking on the footing the front, near end corner of the wall. Scratch a mark in the concrete or drive a nail into it, if you can. Naturally, the block is to be more or less centered on the footing, so the mark must be positioned accordingly. Repeat the operation at the far end. Then using the nail or the help of a friend, snap a chalk line between the two points. This line marks the front edge of the wall.

Now scoop up a generous glob of mortar with a trowel. Slosh the mortar over the surface of the footing near the mark that indicates where the near end of the wall begins. Spread the mortar with the trowel to make a fairly even bed about 24 inches long, 1½ inches thick and 2 inches wider than the width of the block. Carefully lower a corner block onto the mortar. Keep the corner of the block exactly over the corner mark (or alongside the nail), with the long side of the block exactly in line with your chalk mark. With the end of the trowel handle tap and push the block down into the mortar, making it as level as you can by eye. Then, using a spirit level at least 30 inches long, make certain the block is level both lengthwise and across its width. Now repeat this operation at the far end of the footing.

The next step is to run a line from the top front edge of one of the blocks thus installed to the corresponding edge of the other block. Wrap the line around a brick or similar heavy object several times and place the brick near the end of one block with the line a fraction of an inch clear of the front of the block. Repeat this with the other block at the further end of the string. Now move the two blocks, if necessary, until their front sides are aligned with the line and with the chalk mark below the line. If you have had to move either block, make certain it is level in both directions by using the level again.

Next, place a line level in the middle of the line to check the relative elevation of the two blocks. If the difference in height is more than $1/2$ inch, measured by lifting the low end of the string until the bubble is centered, push down on the high block, but make sure that there is still at least $1/4$ inch of mortar underlying it. Or raise the low block by removing it, applying more mortar and repositioning it. If the difference in height is $1/2$ inch or less, keep the difference in mind, and when you lay the next course, make the mortar a fraction thicker at the low end than at the high end. Repeat this with every subsequent course until the ends of the wall are level with one another.

As soon as you have the end blocks properly placed and level with one another, or nearly so, it is time to position the rest of the blocks between them. These middle blocks can be end or stretcher blocks. Let's assume they are stretcher blocks because it is easier in this case to explain how the mortar is applied.

Start by standing the first block up on end (if you have not already prepared a row of blocks this way). Take a gob of mortar on your trowel and throw it on one flange of the up-ended block. If you do this right you will leave a layer of mortar about 2 inches high on the flange of the block. Then with a downward, angular sweep, press the mortar against the end flange with the underside of the trowel the way ice cream is pressed into a cone with a scoop. The pressure should reduce the mortar thickness to about 1 inch. This is called buttering the block. Repeat the operation on the other end flange. Keeping the small holes up, lower the block onto the mortar bed you have previously prepared and gently pressed sideways against the block already in place. Hold the block by its webbing when you do this. When working with 12-inch-thick block it is advisable to have a second person holding the

LAYING
CONCRETE BLOCK

1 Stand the blocks you are going to use near at hand so that you don't have to waste time and energy walking back and forth to lay them.

2 Mark one corner of the course on the footing. Here a plumb bob has been dropped from batter board lines marking the corner.

3 Locate the second corner and snap a chalk line between the two marks.

4 Spread a layer of mortar on the footing, an inch or two thick, and several inches wider and longer than the block. You'll cover your guideline, in part, but this can't be helped.

5 Lay the block on the mortar, the corner against the nail or corner mark. Align the block with the visible part of the guideline.

6 With a spirit level and a hammer, level the block first in one direction, then in the other.

block with you. As you lower the block and move it sideways against the corner block, take care to keep its front side aligned with the line. The block should not touch the string, even lightly, or it will push it out of line. The string establishes both the position of the front of the block and its height. Now with your level and the heel of the trowel level the block as necessary in both directions.

Spread more mortar on the concrete footing and continue to lay blocks, one alongside the other, until you reach about halfway along the wall. Then go to the far end and lay block the same way, working towards the middle of the wall. Make certain the top and front surfaces of all your blocks are a hair's width short of the guiding line and that all the blocks are level.

7 Repeat the procedure at the far end. Position a second block there as accurately as you can.

8 Stretch a line from one block to the other, using a pair of bricks. Then, if necessary, tap the block into alignment. Recheck the block for level in both directions. Repeat the procedure with the other block.

9 Scoop up a gob of mortar on the trowel and lay it on one flange of an up-ended stretcher block. The mortar should be about 2 inches high.

10 Press the mortar into place with the flat of the trowel. The pressure should reduce the mortar to a thickness of about 1 inch. Then butter the other flange the same way.

11 Lay the block in place, sliding it gently against the corner block. Complete the first course in this way.

12 Turn the corner and lay a course at right angles to the first course. Here the end block is being aligned, leaving a space for the last block, which will be cut to fit.

13 The closure block is cut with the blade of the brick hammer by tapping along the "break" line on both sides of the block until it cracks.

14 Butter the ends of one block in place and the opposite ends of the smaller block. Lower the smaller block into the opening.

Joint patching. Until you learn to keep the consistency of the mortar just right and how to make the mortar stick to the ends of the block while you lower it into place, you will find yourself with a lot of open joints. *Do not lift the block out of place again.* There is no need for this. Instead, put a gloved hand between the flanges and behind the open joint; then take some mortar on the back of the trowel and force it into the joint, using your gloved hand to prevent the mortar from getting inside the block. Joints made this way are as strong as those made any other way; they just take longer.

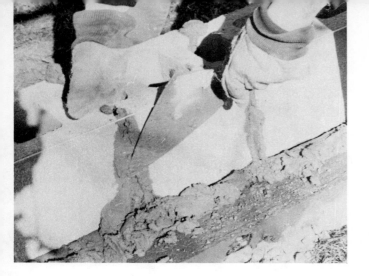

15 Some of the mortar will fall away, but don't lift the closure block. Instead, place a gloved hand behind the joint and force fresh mortar in with the trowel.

16 Place mortar on the face shell of the first-course corner blocks and lay the corner block of the second course in place. Note that the new block overlaps the joints of the lower blocks.

17 Make certain the top block is level in two directions and that it is flush with the lower blocks.

337

18 Build up the corners of the wall, checking that the blocks are flush with those in place.

19 Clean up your work as you go along, using the side of the trowel and scraping the joint surfaces flush with the surface of the blocks.

20 When the mortar has stiffened somewhat, compact the joints with a joint tool.

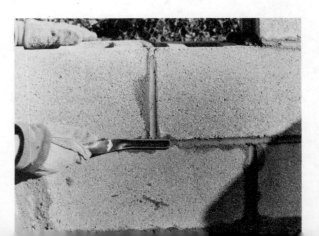

21 Build up the opposite corners using the completed corner as a guide. Then lay the middle blocks.

Closure block. When the space remaining between the blocks you have laid down is the length of one block or less, you are ready for the closure block. This can be a regular block, or a block cut with a brick hammer to 1 inch shorter than the opening. Butter both ends of the closure block generously, lower it into the space, and level it.

Cutting block. When you need to cut block to an exact size you can do so with a power saw and a diamond or abrasive blade. For ordinary purposes, however, concrete block is cut by breaking. This may be done with the blade of a brick hammer or the edge of an ordinary hammer. Lay the block on sand or soft dirt, or hold it up at an angle, resting on one edge. Do not pound on the block while it lies flat on a solid concrete or similar surface. Rap one side of the block fairly hard along the line at which you want it to break. Then turn it over and do the same on the other side. Keep rapping the block along the desired break line first on one side then on the other until it separates into two pieces. It should break cleanly if you do not strike it too hard at one point. If it doesn't break satisfactorily, try another block. That is why you always order a few extra.

Laying down the following courses. Since the first course began with a whole block, the second course must start with a half block so that the joints in succeeding courses are staggered. Start by cutting a corner block fairly evenly along its middle. Then butter the top of the corner block of the first course. Apply the mortar as already described until the edges of the block—and a little more—are buttered to a thickness of about 1 inch.

Now place the half block on top of this mortar, with its finished end flush with the finished end below and its cut end pointing towards the far end of the wall. Push the piece of block down and move it around if necessary until its sides are flush with the block below. Stop pushing downward when the mortar joint is about ½ inch thick—unless you have a reason for wanting a thicker or thinner joint. Using the spirit level and the heel of the trowel level the half block perfectly both along and across the wall, and check also that it is vertical. With the side of the trowel scrape the excess mortar from the sides of the blocks so that you can place your level against a clean surface. Repeat the operation at the far end of the wall. The reason for working first at one end and then at the other is to permit the mortar to stiffen up a bit, which makes the operation easier for beginners.

With the end blocks of the second course in place, the line, which obviously has been moved out of the way in the meanwhile, is strung from these two end blocks just the way it was strung from the two lower blocks. Mortar is now placed along the top sides (not the webs) of the blocks next to the end blocks of the first course, covering a little more than three blocks. Stretcher blocks are buttered as previously described. They are then positioned end to end, just like the blocks of the first course, following the half blocks already placed at the ends of the wall. At this point you have three and a half blocks of the second course at both ends of the wall. You can, if you wish, fill in the remaining space, but most masons prefer to bring the wall ends higher before filling in the middle.

So returning to the near end of the wall, butter all the top surfaces excepting the webs for a distance of a little more than two blocks. Start by placing an end block on top of the end of the two courses of block, its end and sides flush with those of the wall. Scrape the excess mortar away with your trowel edge. Then with the spirit level make certain the block is perfectly level both lengthwise and across, and that it is perpendicular and perfectly flush with the blocks below. Now add a second stretcher block following the top block you have just positioned. Again, make certain it is level and flush. Repeat this operation at the far end of the wall.

At this point you have one complete course of block plus a second and third course consisting respectively of three and two blocks at the corners. You can now go on to position the ends of the fourth course, which must start and end with another half block. The same procedure as before is used, bearing in mind that the higher the wall goes the more careful you must be to make certain you hold the corners perfectly vertical.

Now you can start filling in the inside of the wall. However, you can no longer use a brick to hold the line. In its place you can use a line clamp made for the purpose or do as most masons do: drive a nail into the mortar and tie the line to it.

Going higher. We have described the basic procedure for building a wall four courses high. A higher wall can be constructed the same way; the corners are simply made higher before the middle section is filled in. Or the four or five courses are completed and corners or ends are constructed on top of them. One method is no better than the other; the choice is merely one of convenience.

Precautions. Make it a habit to automatically clean the inside and outside surfaces of the blocks as you work to prevent excess mortar from hardening there. For one thing, this makes the job unsightly. For another, if you hold your level against a blob of mortar the reading will be incorrect.

Do not reuse mortar that has fallen to the ground. If the mortar picks up dirt it will not adhere properly. Keep a constant check on the workability of the mortar. Add a little water as necessary and dump the mortar when it is more than two hours old.

Keep your tools and mortar board or pan clean. As soon as you have used up a batch of mortar, wash the trowel, mortar board, and spirit level. You may be unpleasantly surprised by how tenaciously dried mortar can hold on to metal.

Keep checking the length of the wall for pitch as you work. It is very easy to start going uphill or down without noticing it.

Watch the closures of the second and following courses. You don't want one to be directly above another, if this is avoidable.

Watch your hands. Keep them dry and clean or they will quickly become sore and then blister. Cotton gloves help a little, although they reduce your grip and so make working more difficult. However, you should always use gloves when moving block and then take them off to use the trowel and level.

Don't work from ground level on a wall that is above your shoulders. It is too difficult to lift blocks this high. Erect a stable scaffold and work from it. Don't use a ladder; it is too difficult and too dangerous.

Get help on 12- or even 10-inch block if you are unused to the work. It is difficult enough for a beginner to lower a block he can handle into position without tottering under its weight.

Wherever the local codes permit, use lightweight block. It costs a little more but is almost as strong as regular block.

Keep your block dry because it is easier to lift dry block than wet and because, as explained, you can use mortar of a more suitable consistency.

FINISHING THE JOINTS

This operation is also called pointing and joint tooling. Where appearance is not important, simply drag the point of the trowel down the lengths of all the joints. This forces the mortar more deeply into the joints, compresses the mortar and makes the surface of the joint fairly smooth. Both sides of the wall should be done if possible. Then with the side of the trowel the surface of the block is scraped free of any mortar that was pushed out of the joints or adhered after falling off the trowel.

For more even and attractive joints, use any of the many joint tools available. Some produce a V indentation and others a half-round one. You can make the latter shape if you wish by using a bent length of ½-inch tubing. The effect of all of these tools on the mortar is exactly the same; only the surface appearance is different. If there is an open spot as you point, push some fresh mortar in and run over it with the joint tool.

Concave joints are tooled with a ⅝-inch round bar.

V-joints are tooled with a ½-inch square bar.

After horizontal joints have been tooled, verticals are tooled with an S-shaped tool.

Mortar burrs are trimmed with the trowel. *Photos courtesy Portland Cement Assoc.*

CONCRETE-BLOCK FOUNDATIONS

If you have gone so far as to purchase a lot, excavate a hole and construct a concrete footing, you might as well continue right on and build the foundation. Although there is a lot of work involved, in the sense that a considerable quantity of material has to be moved and a large number of blocks have to be laid, the work itself is neither technically nor physically difficult. The job does not have to be perfect. It can be out of true by an inch or so and the final structure can still be beautiful. As for strength, many masons don't believe it is really necessary to use mortar on 10- and 12-inch-wide block; the weight of the blocks and the building on top will hold it all in place (except in earthquake and hurricane territory). Besides, neatness, the mark of an experienced mason, doesn't count as far as the strength of a concrete-block wall is concerned; blocks laid a little crooked are just as strong as blocks laid perfectly straight. So if you are hesitating from lack of experience, hesitate no longer. You'll be a fair to middling mason by the time you have finished.

As for the labor involved, it is considerable when totaled, but it doesn't have to be accomplished at any particular pace. On average, two experienced men—a mason and a helper—can erect a 30-by-40-foot concrete-block foundation, fourteen courses high, in approximately three, thirty-five-hour weeks. So figure a pleasant summer's work for one man. (Because of the time spent walking back and forth, two men can do the job in about a third of the time it would take one man.) The word pleasant is not used lightly; working at your own pace, masonry is as enjoyable as gardening or any similar outdoor hobby.

ESTIMATING MATERIALS

Blocks. Figure the number of blocks you will need for the exterior and interior walls of your foundation using the method described in the preceding chapter. Don't over order by very much as comparatively few blocks will be crushed in delivery. An extra seventy-five blocks for the average house foundation should be plenty. Don't forget that you may need solids to cap the foundation and anchor bolts to tie down the sill plates of the house to be erected on top. The solids are simply laid flat on top of the foundation wall and all the bearing walls within the foundation perimeter. To estimate the number of solids you will need, divide the length of the walls to be capped, in inches, by the length of the blocks you plan to use.

Sand and cement. Follow the suggestions made in the preceding chapter to estimate the quantity of sand and cement you will need. If you are going to work by yourself and expect the job to be a long one, spend the extra money and have the cement delivered in small loads. This will reduce the possibility of the cement hardening in the bags because of absorbed moisture.

Don't forget the sand and cement and possibly lime you will need for parging, which is plastering the outside of the foundation wall with mortar to make it waterproof and hide the joints above grade.

Generally, the mortar coating is $1/2$ inch thick. To estimate the quantity of mortar necessary, determine the surface area of the foundation wall in square feet and divide by 24 to get the number or cubic feet of mortar you will need, then increase the figure by 20 percent to take care of the mortar you will drop as you work.

For example, a foundation 30 by 40 feet and 10 feet high will have a surface area of 1,400 square feet. Dividing by 24 (if the coating is to be only $1/2$ inch thick) gives 58 cubic feet. Adding 20 percent brings the number up to 67 cubic feet.

You can use any of the mortars suggested in the table in the preceding chapter. The mortar with the most lime is the easiest to apply, since lime makes mortar slick and sticky. Type M, however, is the strongest and least susceptible to water penetration.

One way to lock concrete walls is with strap-iron tie bars held in place with mortar. If manufactured ties are not available, make them yourself by bending lengths of ⅛-by-1-inch strap iron. The bent ends should be at least 2 inches long.

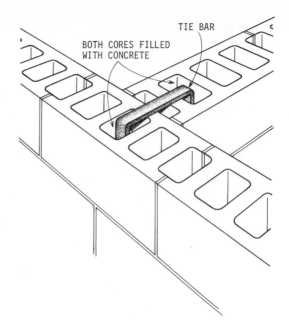

BOTH CORES FILLED WITH CONCRETE

TIE BAR

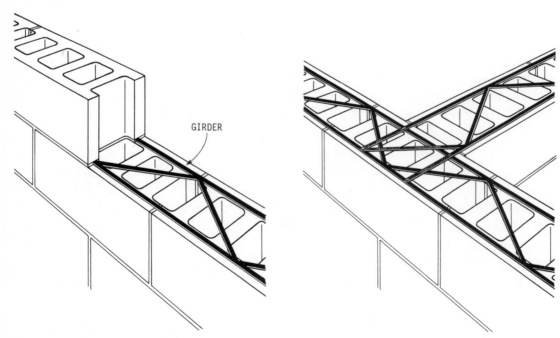

GIRDER

When a girder-type reinforcement is used to strenghten a wall, it is placed between courses of block. The block is laid with mortar in normal fashion (left). To reinforce two walls that meet at a right angle, one girder is laid on top of the other (right).

346

REINFORCEMENTS

Most foundation walls are constructed without the use of any steel to strengthen the joints. However, when the footing rests on a variety of supports, as for example on bedrock and soil, fill and virgin soil, or any other combination that may settle differently, reinforcements are advisable.

The wall areas most in need of such reinforcement are the corners and the joints between an interior and an exterior wall, the wall above the spot where the footing support changes, and the wall beneath a large window.

Four general types of reinforcement are used. One is a hooked strap of steel that is laid across two blocks and clamps them together with its bent ends. Another consists of expanded metal cut into long strips for lengthwise use on a wall, into angles for corners, and into Ts for interior to exterior wall joints. A third type, which looks like a wire girder, consists of two parallel steel rods joined by a zigzag welded rod. The last and least used type consists of standard steel reinforcement bars, generally $1/4$ inch thick.

Reinforcements can be cut with a bolt cutter, a hacksaw, or large tin snips. They are easily installed. To install mesh- or girder-type reinforcements, the tops of the blocks are buttered with mortar, the mesh or girder — $1\frac{1}{2}$ to 2 inches narrower than the block — is laid in the mortar and the next course of blocks laid on top. To make a corner, the ends of the two girders are placed at right angles, one on top of the other. To make a T joint, the end of one girder or strip of mesh is placed over the middle section of another. To join one strip or girder end to another, they are overlapped, and the mortar locks them together.

The strap reinforcement requires that the holes into which the strap ends fit be filled with mortar. The next course of blocks is laid as usual.

The bars are placed on top of the mortar just like the strips and girders. When the next course of blocks is positioned, the bars are automatically pressed into place. To reinforce corners and T joints, the bars must be bent to shape.

PLANNING MATERIAL DELIVERY

Bear in mind that the closer you can get the material to the job, the less you will have to haul. Since the average foundation requires close to 2,000 blocks, the total weight — block, cement, and sand — that has to be moved into position can amount to

60 or 70 tons. That's a lot of hauling to do by yourself with a wheelbarrow or the garden-type barrow used for carrying block. So think carefully about getting the truck as close to the foundation as possible. The best place for the block is usually spaced around the inside of the footing, about 6 to 7 feet away to give you room to work and to erect a scaffold, should you need one.

Modern, block-delivery trucks have long crane arms that can reach about 10 feet or more so that it is possible in many locations for the truck to place pallets where you want them without the truck entering the excavation. But if this is not possible, give serious thought to covering part of the footing with a foot of soil so that the truck can ride safely over it and drop all or some of the block inside the foundation footing perimeter. (Be certain to wash the soil from on top of the footing before you lay the block.)

The sand and cement are best placed outside the excavation to reduce their exposure to moisture and preclude the possibility of flooding. The cement must be raised above the ground—planks set on top of blocks can be used—and completely covered with plastic sheeting. The blocks too should be protected from the rain; but it is enough if you cover the tops of the pallet-loads. There is no harm in letting the sand get wet; you just add less water.

BUILDING THE FOUNDATION WALL

Start by returning the four lines to their original positions on the batter boards. (For simplicity, we are assuming you are going to work on a rectangular foundation without any openings; they are discussed later on.) Call the side nearest the road the front side; the side to your left, with your back to the road, the left side. That makes the side to your right the right one and that farthest from the road the rear one.

Next, carefully mark the outside corners of the foundation by dropping a plumb bob from the four points where the lines stretched between the batter boards cross one another. Drive a nail into the concrete, if you can, or scratch a mark where the bob's point indicates. Then snap a chalk line between the four nails or marks. This gives you the outline of the foundation.

At the front left corner spread a layer of mortar along the front footing, from left to right for a distance of two or three block lengths. Lay an end block on the mortar, its finished end flush with the left-side chalk mark and its length flush with the chalk mark indicating the front of the foundation. If the mortar has covered the chalk line —which it may well do—use your eye to align the block. Then, as previously described, level the block using the trowel and a spirit level. Next, lay a stretcher block to the right of the first block. Align it with the chalk line on the front footing and level it, top flush with the corner block.

Now, at the front right corner lay down a bed of mortar from the front chalk mark towards the rear of the building. Lay down another end block with its finished end towards the front of the building and its length along the right-hand side of the building. Make the block level and align it with the right-side chalk mark. Check that the corner of this block, like that of the block at the other corner, is exactly in line with the corner nails or marks. Drop the plumb bob again to make certain.

A seven-course corner reinforced with steel girders at every other course. Mason is stretching a line to another erected corner preparatory to laying block between the two corners.

Repeat this process at the two remaining corners. When you have done so, lay up some more block at the first corner. Build the corner up three courses. Take your time and make certain the corner remains exactly where it should be and that all the blocks are flush with one another and level. Note that you must alternate the direction of the end blocks at the corners so that they overlap one another, thus making a strong joint.

After you have erected all four corners, three courses high, it is time to check your progress. Drop the bob at the four corners. Measure diagonally from corner to corner to see whether or not you are holding the foundation square. Then measure width and length at the four corners to make certain your foundation is not growing larger or smaller. If it is smaller this isn't too great a problem, since you can always make the plaster coating thicker to make up the difference; but if it is larger you can run into serious difficulties. Then check for level. Use a transit, water tube, or line level as discussed in Chapter 16 to make certain the heights of the four corners are the same. Try for perfection, but if you end up with no more than an inch difference in 40 feet or so, you are fine.

When the transit cannot be conveniently set up in the middle of a foundation, it can be placed outside the perimeter, as here. After leveling the transit, the mason is sighting the ruler held by his helper, who will then move it around the wall for further sightings. In this way, they check the elevation of each side throughout construction so that corrections can be made by increasing or decreasing the thickness of mortar joints between the blocks.

Correct for errors in foundation width and length by moving the succeeding blocks inwards or outwards as required ¼ inch at a time. More will be too obvious. Correct for height errors with similar upward or downward modifications. Naturally, when you have the corners up, you can fill in between them whenever you wish. Add the reinforcements, if you are using any, as you go.

CAPPING THE FOUNDATION WALL

When you have laid down the desired number of courses, your wall's top course will consists of a row of block with open holes on top. These holes must be filled and a number of anchor bolts must be installed, assuming a frame structure is going on top.

There are several ways to "cap" the foundation, as filling or covering the holes is called. The choice of methods will depend on your building department. In this writer's opinion, one is no better than another. Pick the one that appeals to you or is permitted.

The easiest consists of filling the holes with concrete or mortar, but in order to keep from filling the entire wall, you have to provide some kind of stop inside or under the top row of holes. One way is to throw odd stones and pieces of broken block down the holes, then fill them with concrete or mortar. Another is to stuff wet

Some codes require a foundation wall to be topped by 4-inch solid blocks. You have to butter webs and edges of top course to lay the solids. Anchor bolts are installed between the blocks.

newspaper down the holes and then fill the top 5 or 6 inches with concrete or mortar. An easier way consists of placing a strip of tar paper, just wide enough to cover the holes, along the penultimate course. Butter the block edges, alongside the tar paper, just as you normally would. Then lay the top course of block. With the tar paper in place, the cement cannot fall any farther than the depth of one block.

Where required, solids are laid on top of the final course in the usual way. However, since the solids are only 7⅝ inches wide, they may not cover the full width of the underlying blocks, which probably will be 8 inches. Therefore you have to butter the webs of the blocks as well as the edges. Keep the cap blocks flush with the outside of the wall.

At the same time that the foundation wall is capped, the anchor bolts are installed between the blocks, head down, threaded ends up and properly spaced so that they will not interfere with the floor joists. If necessary, break a solid in two to permit the bolt to be positioned correctly.

DOORS AND WINDOWS

Up until now we have assumed the foundation will be rectangular with no fenestrations; but that is, of course, unusual. Most foundations have doors and windows. Here is how to handle them.

Doors. The easiest method is to position the door frame where you want it as soon as you have erected the two corners of the wall in which the door will be set. Raise the door frame to the correct height with bricks and stones — remember, you are going to pour a concrete floor as soon as the foundation is completed. Use stakes and lumber to hold the door frame perfectly vertical and immobile in its final and correct position. You now continue building the wall on either side of it using the edge of the frame as the starting point for a wall end and working towards the corner that you have already erected.

If you are installing a simple, flat wood frame, such as is common for a cellar or basement entrance, you can start the wall adjoining the door frame with an end block that just touched the wood, but does not press against it. Later, concrete nails can be driven through the wood and into the block. The joint between the frame and the blocks can be sealed by caulking it. On the other hand, if you are installing a more complex frame of wood or metal, you may find it more convenient to enclose the

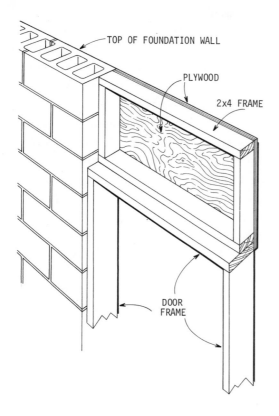

TOP OF FOUNDATION WALL

PLYWOOD

2x4 FRAME

DOOR
FRAME

One way to handle a doorway is to leave an opening in the wall from the ground to the top course. The door frame is installed in the opening and the space above is filled with another frame covered with plywood.

frame with jambs. Note that wood frames are always positioned a fraction of an inch above the concrete floor. The bottoms of metal frames may be covered with concrete. The door sill can be put into position before pouring the floor, in which case the concrete will lock it in place. Or it can be positioned after the concrete is poured, assuming the floor runs through the doorway and beyond. In any event, make certain you allow for the concrete floor and the sill, if used, when you position the frame and before you lay block up alongside it.

If you are using a wood frame or buck, as it is often called, you can eliminate the need for a metal or stone lintel by using a wood frame and plywood to cover the space between the top of the frame and the sill overhead. If you are using a metal frame, you will probably want to install a steel or stone lintel, a feature that is discussed after windows.

Windows. These are treated exactly like doors, except, of course, that they are placed later in the building operation. For example, assuming the bottom of the window is to be positioned a few inches above the sixth course, you go right ahead and lay six courses. Then you place the window frame in position, using stakes or similar aids to hold it there. If there is to be a sill, it is positioned now. Sills, just like block, are held in place by mortar. If you are not going to use a precast sill, you can make one by shaping one out of mortar. This is easily done in a case where the top edges of the window are flush with the top of the foundation. Leave an opening that is 2 inches wider than the window frame and 2 inches higher. Wait until construction has been started on the house frame and the sill plates are in place. Then nail the top of the window — which is made for this type of installation — to the sill plate. Lock the window against the block with wedges of wood so that its bottom cannot move. Now apply cement to the edges of the block with a steel trowel, forming a smooth angle from the bottom edge of the window to the edges of the blocks on the sides and bottom. Remove the wedges as you go. Do the same along the sides of the window frame.

Other frames will require different treatment. Just be certain you examine the frames and measure them up for the necessary clearance before you begin laying the block.

One side of an opening left in the foundation for a window. Note use of corner block and short block in second from topmost course. Since the opening is comparatively narrow, the code in this area permits a window without a lintel.

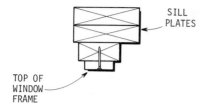

SILL
PLATES

TOP OF
WINDOW
FRAME

Method of installing a window. Top of frame is nailed to bottom of sill plate (left). The bottom of the frame is locked against the block with a wedge. Then mortar is applied to the bottom of the frame and it is trimmed flush with the block (below).

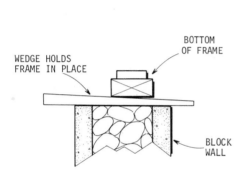

WEDGE HOLDS
FRAME IN PLACE

BOTTOM
OF FRAME

BLOCK
WALL

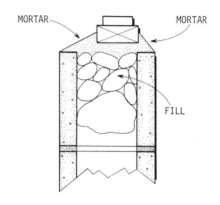

MORTAR

MORTAR

FILL

Lintels. A lintel is a kind of metal or stone bridge placed over a window or door for the purpose of carrying a load that cannot safely be permitted to rest on top of the door or window frame. The easiest type to use is made of reinforced concrete and is just as thick as the width of the block. The wall adjacent to the window or door frame is constructed to a height flush with the top of the frame. Then the lintel is placed above the frame with its ends resting on the block. A little mortar is used under it to hold it in place. Then, whatever additional blocks are necessary are laid on top of the lintel as usual. The only precaution here is to make certain you get at least 4 inches of each end of the lintel on the supports. (Some local codes require more.)

Two types of metal lintels are used with concrete blocks. One has a T-shaped cross section. These lintels are placed flat side down, and concrete brick or whatever fits is cemented in place, front and back, to hide the metal. The other type is an angle iron. It is generally placed with one side of the metal exposed and the other support-

Method of installing a doorway in a block foundation with a lintel. The frame is braced while the wall is built. When the blocks are flush with the top of the frame, the lintel is laid in place. Small blocks or bricks are used to fill the space between the course. One method of fastening the door frame to the block is shown below.

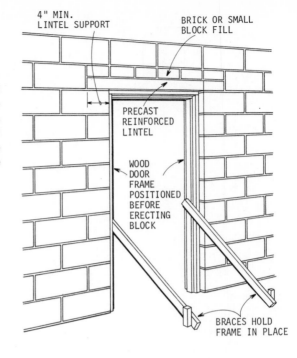

4" MIN. LINTEL SUPPORT

BRICK OR SMALL BLOCK FILL

PRECAST REINFORCED LINTEL

WOOD DOOR FRAME POSITIONED BEFORE ERECTING BLOCK

BRACES HOLD FRAME IN PLACE

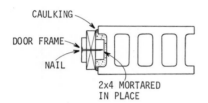

CAULKING

DOOR FRAME

NAIL

2x4 MORTARED IN PLACE

ing the block. Any space between the window frame and the metal lintel and the block at the sides is sealed with caulking.

In all cases, make certain the lintels are approved by the building inspector before you install them. If you are outside a code area, the supply yard can usually tell you what is safe in regard to load and span, if your print doesn't specify the lintels.

OTHER FEATURES

Girder supports. If your building has a central girder that is carried by the exterior walls, pockets (spaces) must be left in the walls to accommodate the girder ends. Here again we enter the realm of bureaucratic confusion that costs builders and home buyers so much time and money. Some codes permit you to merely cement the supporting block surface smooth, some insist you lay down a solid or concrete brick beneath the girder end, and others insist a metal plate be positioned here. In any case, unless the code or your plan specifies otherwise, fill the block and give the girder end 6 inches of supporting surface; that will be plenty.

Interior walls. Load-bearing interior walls must be carried on footings equal in size and depth to that used for the exterior walls. These interior walls are usually constructed or poured at the same time as the exterior walls. They must be capped and provided with foundation bolts, again just like the exterior walls. The junction between the walls forms a T joint, which is constructed by overlapping the blocks that form the joint.

Non-load-bearing interior walls can be joined to exterior walls by constructing both simultaneously and overlapping the blocks and also by means of metal reinforcements used without overlapping the blocks. Since the walls do not carry any load, they can also be constructed on the concrete cellar floor and joined to the exterior walls with mortar.

Chimney foundations. The foundation of an exterior chimney consists of a comparatively small rectangle or square of block adjoining the inside or outside of the foundation wall. Like an interior bearing wall the chimney must be carried on a footing. The chimney foundation is joined to the exterior wall by overlapping the blocks at the joints.

Pilasters. A pilaster is a very short wall placed at right angles to another wall, usually for one of two reasons: to provide lateral support and to carry the end of a girder. When a house butts up against the side of a hill that is known to be wet and contain a lot of clay, making earth movement a strong possibility, the foundation wall up against the dirt may be reinforced with two or three pilasters. In such cases the pilasters may extend two or three feet into the cellar and rest upon footings equal in thickness and depth to the perimeter footings. The joint between the pilaster and the exterior wall is made by overlapping the blocks. When a heavily loaded girder has to be supported by a thin exterior wall, the wall is thickened at the point of support by a pilaster.

WATERPROOFING THE FOUNDATION

Even if there is no water in the excavation indicating the presence of a water problem, it is customary to waterproof the foundation to keep its insides dry and also to hide the block above grade.

Plastering. Start by plastering, or parging as it is sometimes called, the outside of the concrete-block foundation wall. (This is usually not done with poured walls.)

PILASTER

Pilaster footing should be poured with the foundation wall footing and the pilaster joined to the wall by overlapping the courses.

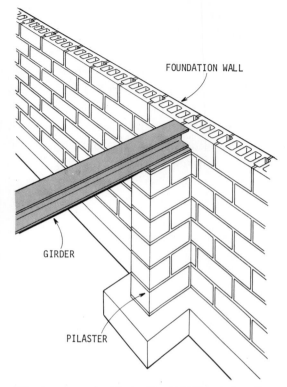

FOUNDATION WALL

GIRDER

PILASTER

Pilasters are used to buttress basement walls and to provide support for girders.

Use any of the mortars suggested in the previous chapter. Type M, however, will provide maximum water protection. You will need a hawk, a steel trowel, and a mortar board.

Mix your mortar a little on the wet side, but stiff enough to be piled into a cone, and place it on the mortar board. Hold the hawk, handle down in one hand, with the top of the sheet of metal to one side and below the mortar board. With the trowel push some mortar off the mortar board and onto the hawk. Then tilt the hawk and swing the trowel upwards just as a little mortar slides off; and with a backhand, upward swing of the trowel spread the mortar over the block. When you have spread all the mortar that was on your hawk, smooth it with a pressing, circular motion. Generally a layer $\frac{1}{2}$ inch thick is used. The thickness need not be exact, but it should be more than $\frac{3}{8}$ inch thick at the thinnest point. When you have plastered a

Apply plaster (mortar) with a steel trowel using an upward stroke.

Use trowel-consistency asphalt and apply it with an old or an inexpensive trowel to seal the foundation against water.

section of wall, cover the exposed surface of the footing and the bottom edges of the wall with mortar. Make this surface one continuous curve or sweep upwards. Again, you want about ½ inch of mortar on top of the footing; more won't do any harm. When finished, the mortar coat should be one smooth, continuous layer from the bottom to the top of the foundation wall. Give the mortar a couple of days to harden thoroughly. If the weather is hot and dry, moisten it with a light spray of water to speed curing.

When the mortar coating is hard and dry it is covered with a layer of asphalt. This can be applied with a cheap metal trowel that is later discarded. Working directly from the can, apply the asphalt as evenly as you can in a layer about ⅛ inch thick. Be certain to get the trowel-type asphalt and not the liquid type.

When you expect lots of water pressure, cover the asphalt with a layer of 30-pound roofing paper. The asphalt will hold it in place.

For moderate to bad water conditions it is customary to cover the asphalt with a layer of 30-pound roofing paper. This can be done simultaneously with the application of the asphalt, which acts as a sort of glue. Start by pressing the bottom edge of the paper to the asphalt-covered wall and then press upwards. Use sufficient paper to reach above the finished grade. Cut it with a hook knife and make joints by overlapping the pieces. Go around the corners with at least 2 feet to spare. The paper should curve outwards onto the top of the footing but need not go below it.

DRAINING THE FOUNDATION

This is generally done after the foundation wall has been plastered and tarred, but it can be done before if this is more convenient.

Start by digging a 1-foot-wide trench alongside the footings. Let the footing itself form one side of the trench and make the bottom of the trench level with the bottom of the footing. Fill the trench with a 3-inch-thick layer of gravel or crushed stone. Place 4-inch plastic or bituminous fiber drainpipe in the trench, with the holes at the sides of the pipe. Lift the pipe where necessary to give it a ½-inch to the foot pitch in the direction or directions you wish the water to flow. Use stones to hold the

Drainpipe positioned alongside footing and partially covered with crushed stone. Note position of drain holes. When the job is completed, the entire pipe will be covered with stone and the stones will be covered with tar paper.

pipe up, and a spirit level to judge the pitch. Next, gently surround the pipe with crushed stones or gravel. After checking the pitch once again, cover the pipe with at least 6 inches of crushed stone or gravel; more if water conditions are bad. Level the surface of the stones with a rake. Then completely cover the stones with a layer of tar paper. The paper is in turn gently covered with at least a foot of soil, after which you can complete the backfilling with a machine. If you backfill without the protective layer of soil, there is a good chance the tar paper will be ripped. This permits the soil to plug up the spaces between the stones, thus reducing water flow into the drainpipe.

CONCRETE-BLOCK BUILDINGS

Concrete block is probably the most practical and least expensive building material presently available. It has no peers for small-building construction. It is strong, fireproof, rot-proof, insect-proof and for all practical purposes, ageless. Concrete block has only one drawback. Its insulating value is nil, but that can be remedied by applying a layer of insulation to the inside walls of the building and by making block walls hollow and filling the space between the blocks with insulation.

The construction of a small concrete-block building, a garage or toolshed, for example, is no more difficult than assembling blocks for foundation walls or making footings and concrete slabs. All the techniques necessary have already been discussed in previous chapters. The information given there and in this chapter by no means exhausts the subject of constructing small concrete buildings. There are many other ways to do the job, but our approach is the simplest and requires the least equipment and experience.

BLOCK SIZE

Unless local codes specify otherwise, you can use 6-inch block for sheds and similar buildings that do not have above-ground walls higher than 8 feet or roof spans of more than 10 feet. Use 8-inch blocks for buildings which have walls of more than 8 feet in height, but which are not two-story buildings and do not have roof spans greater than 20 feet. Use 10-inch blocks for two-story buildings with roof spans of up to 25 feet.

FOUNDATIONS

Concrete-block buildings can be constructed on either a slab or a perimeter foundation. The choice of footing depends on building size, the nature of the soil, and the weather.

Slab foundations. When the building is small—let us say it is a chicken coop 6 feet high and 8 feet wide made of 6-inch blocks—and there is no frost to speak of, the blocks can be laid directly on a 4-inch concrete slab. The slab can be constructed and poured in exactly the same way as a driveway or patio slab. Its surface can be floated or troweled as you wish. If you make the slab larger than the building, a portion of it can serve as a sort of stoop.

When the building is larger, but not large enough to require 8-inch block, make the slab 6 inches thick and construct the block walls directly on top of it.

When the building requires 8-inch block it is best to reinforce the edge of the slab that supports the block. Follow the suggestions made in Chapter 18 and dig a perimeter trench 1 foot wide and at least 2 feet deep before you pour the slab, which need be no more than 4 inches thick. When there is frost to contend with, the bottom of the trench must go down below the frost line. The form is then constructed along the edges of the trench so that the foundation and the slab can be poured at one time. In other words, if the foundation wall is 2 feet deep and the slab is 4 inches thick, the total thickness of the slab along its perimeter will be 2 feet 4 inches.

When the span of the roof of a single-story building is such as to require 10-inch block, make the poured foundation wall and footing under the slab's edge 18 inches thick.

When constructing a two-story building, which of course requires 10-inch block, make the poured foundation wall 24 inches thick.

These suggested footing thicknesses are based on the load-carrying ability of soft clay. Where a harder soil is uncovered while trenching, foundation wall thickness can be reduced, but should never be less than 12 inches. The bottoms of these footings must always be below the frost line, unless you have reached bedrock, in which case you pour directly onto the stone.

Perimeter foundations. This type of support consists of a poured concrete footing on which concrete blocks are laid. The technique and procedure are exactly the same as those described in Chapters 17 and 24. The footing itself should be twice as wide as the block, and its thickness should be equal to half its width. Thus when using 10-inch block, you require a 20-inch-wide footing that is 10 inches thick.

Construct the footing and the foundation as already suggested exactly as if you were going to construct a frame building on top. Quit laying block when your top course is as high as you want the floor of the building to be above grade, Generally, this will be about 12 inches above the finished grade, which is the ground level after you have done the landscaping. Do not cap the blocks. The foundation you have constructed serves as the form for the concrete floor you will pour. The next step is to plaster the outside of the concrete blocks with a mixture of mortar cement and sand. This is done for appearance only and is unnecessary if utility is the only thing you have in mind.

POURING AND FINISHING THE FLOOR

Start by cleaning the top surface of the foundation wall. Remove all chips and mortar splatter. Smooth the top of the wall with a hammer. Measure from one diagonal corner to the other to make certain the foundation is square. Measure from side to side to check on its dimensions. If these are not correct, now is the time to rectify them.

Next, stretch a mason's line from one wall to the other and hang a line level in the center, to check whether or not the sides of the foundation are level. A discrepancy greater than 1 inch in 30 feet should be corrected. A lesser error can be ignored. No one will notice it. Correct by adding a layer of mortar on top of the walls or wall sections that are low. Make the edges of the added mortar flush with the surfaces of the foundation wall.

Fill the space between the inside surface of the foundation wall and the trench with crushed stone or gravel to within 4 inches of the top of the wall. Then level all the stones with a tamper. Next cover the stones with a layer of 6-mil plastic sheets, stretching the sheets from wall to wall and overlapping them a foot or more where necessary. If plastic sheeting is not available, use any other suitable moisture barrier.

When you have done this, lay a number of 4x4s across the top of the barrier.

Make certain their top surfaces are level with that of the foundation wall. The 4x4s will act as your guides. Now you can pour, screed, and finish the concrete floor. Use a 1 : 2¼ : 3 mix. Then pull out the guides and fill. Finish and cure as usual.

CONSTRUCTING THE WALLS

With the foundation—whether a slab or perimeter one—and the floor completed, you can erect the block walls as previously explained. However, there is one detail that must be considered at this time: how you plan to finish the exterior walls. If you are simply going to clean and tool the block joints, you can proceed forthwith. If you are planning to stucco the exterior of the blocks, it is advisable to install flashing beneath the first course of blocks to be laid above the level of the floor slab. The flashing keeps rain water from working its way up under the lower edges of the stucco and loosening it.

Actual block sizes (width × height × length), inches	For 100 square feet of wall			Per 100 blocks
	Nominal wall thickness, inches	Number of blocks	Mortar, cubic feet	Mortar, cubic feet
3⅝ × 3⅝ × 15⅝	4	225	5	2.5
5⅝ × 3⅝ × 15⅝	6	225	5	2.5
7⅝ × 3⅝ × 15⅝	8	225	5	2.5
3¾ × 5 × 11¾	4	221	4	2.3
5¾ × 5 × 11¾	6	221	4	2.3
7¾ × 5 × 11¾	8	221	4	2.3
3⅝ × 7⅝ × 15⅝	4	112.5	3	3
5⅝ × 7⅝ × 15⅝	6	112.5	3	3
7⅝ × 7⅝ × 15⅝	8	112.5	3	3
11⅝ × 7⅝ × 15⅝	12	112.5	3	3

Quantities of block and mortar needed for concrete-block walls.
The table is based on use of ⅜-inch mortar joints with mortar spread over the face shell only.
Mortar quantities suggested include a 20-percent allowance for waste.

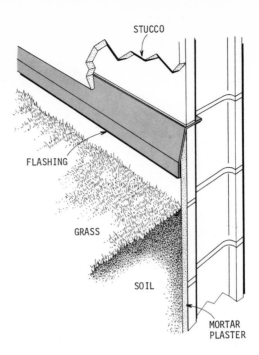

STUCCO

FLASHING

GRASS

SOIL

MORTAR
PLASTER

If you are planning to stucco the exterior of the block wall, it is advisable to install flashing beneath the lower edge of the stucco.

For flashing, lay a 6-inch-wide aluminum sheet on the wall, allowing about 4½ inches to protrude along the exterior edge, as shown in the accompanying drawing. Use bricks or stones to hold the sheet in place while you lay block on top. When the mortar has set hard, bend the protruding sheet downward.

Although the block-laying procedures are exactly the same as when making a foundation wall, the constraints are greater when working above ground. For one thing the blocks will be visible unless you stucco them. For another, it is most important to keep the walls and corners perfectly vertical.

Windows, doors, and other fenestrations are installed as already suggested. One trick that can save you difficulty when you do not have a door or window frame on hand and yet want to continue laying block is to construct a "buck" of 2x6s with the same outside dimensions. By using the buck in place of the missing frame you make certain you hold the opening accurately to the correct dimensions.

CONSTRUCTING A SECOND FLOOR

The second floor is made of wood joists, their ends carried by the block walls. This can be accomplished in any of several ways.

The ends of the joists can be carried by metal hangers made for the purpose. The hangers hook onto the block wall.

366

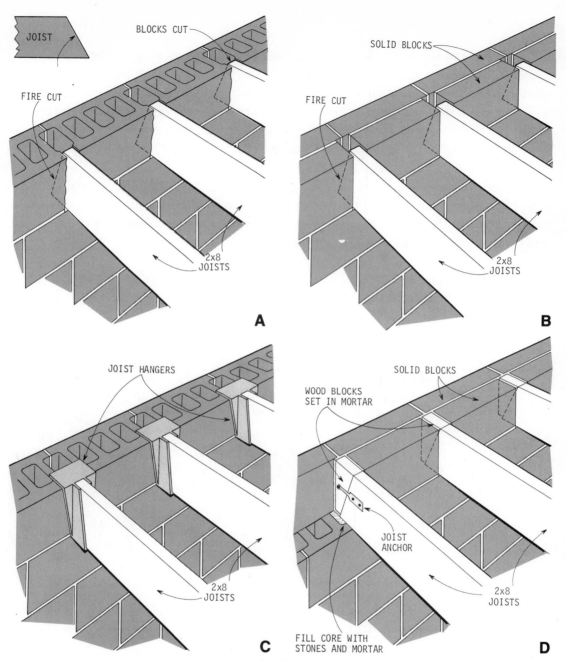

Several ways floor joists can be carried by block walls. The end of the joist must be cut at an angle; this is called a fire cut. (*A*) The ends of the stretcher blocks can be cut to accommodate the joist ends. (*B*) Special joist blocks can be used to provide the needed supporting pockets. (*C*) Hangers can be used to support the joists. (*D*) Anchors can be used to support the joists.

The ends of the joists can be fitted into openings or pockets, as they are called, made in the block wall. To provide the pockets you can cut the facing ends of two stretcher blocks or use special "joist" blocks. When the joist fits into a pocket in the wall the end of the joist must be given a "fire" cut. Very simply, the end of the joist is cut at an angle so that in the event of a fire, the beveled-end joists can fall free without prying the wall block loose. In some communities it is also necessary to tie the joist to the masonry with a joist anchor.

When the floor span is too long for economical joists and a girder is required, the ends of the girder can be inserted into pockets in the wall or the girder can be carried on pilasters. In the latter case the pilaster must be made integral with the wall and must extend all the way down to its own footing.

Supporting surface. When a joist end is fitted into a pocket, there should be at least 4 inches of block beneath the end of the joist, and whatever core openings ap-

Exterior views of roof joists supported on three, on-edge 2x8s. The wall has been capped by 2-inch solid blocks. The wood shingle is placed there as a shim. Later the space between the 2x8s and the tops of the blocks will be slushed with mortar and the shim removed. The 2x8s form a lintel which supports the roof load.

pear beneath the joist end should be filled with mortar. The end of a girder should be supported by 6 inches of masonry. In some codes it is sufficient to merely fill the block cores beneath with mortar. In other codes it is necessary to use one or more 6-inch solid blocks beneath the ends of the girder, be it of wood or metal.

ATTACHING THE RAFTERS

The top of the building walls may be treated exactly like the top of a foundation wall designed to carry a wood frame building. That is, anchor bolts are set in mortar and 2x6 plates installed. The rafters and ceiling joists are then nailed to the plates in the usual manner.

An alternative method is to use three 2x8s on edge in place of a single 2x6 sill. The advantage of this method is that you eliminate the need for reenforcements above door and window openings. The accompanying photos show this method.

INSULATION

Block walls can be insulated by fastening furring strips to the inside of the walls and placing batts of insulation between the strips. Sheetrock can then be nailed across the furring strips and taped and finished in the usual way. To use thick insulation you have to use thick furring strips. As many codes now specify 4-inch-thick wall insulation and 6-inch-thick ceiling insulation, you will have to use 2x4s to secure the necessary space. In such cases you may find it easier to build an adjoining frame wall of 2x4s, nail the plates to floor and ceiling, and fill with insulation.

Block walls can be self-insulating if they are made hollow. This is done by using two 4-inch blocks spaced a certain distance apart rather than a single 8-inch block. The blocks are kept apart and locked to one another by metal ties. The space between the blocks is later filled with loose insulation poured into place.

Obviously the decision to insulate the building must be made early. Not only must you allow for the loss of interior floor space, you must also provide window and door frames of suitable width. Another possibility is to use lumber of suitable width to frame the window and door openings and then insert a standard unit in your frame.

FINISHING THE WALLS

Interior and exterior walls on utility buildings can be left as they are after cleaning and tooling the joints. The interior walls of cabanas and similar warm-weather habitations can be finished by plastering. For this you can use a high-lime cement mixture applied with a trowel and finished with a steel trowel or float. You can also use a brown undercoat, which can be applied the same way and dries a pale brown. Or you can do a standard lime-plaster job consisting of three coats and finished perfectly smooth.

Interior walls can also be finished with Sheetrock nailed to 1-inch furring strips. The walls are then taped and painted.

Exterior walls can be finished by covering them with a veneer of brick or stone, or even a layer of wood nailed to furring strips. They can also be finished with stucco, which is a cement-sand mixture that is applied with a trowel. Its surface can be smooth or gently textured, and its color can range from offwhite through a number of pastel shades.

To cover interior walls with Sheetrock, furring strips must be fastened to the block walls. Insulation material goes between the strips.

Preparing and applying a stucco finish. Prepare the surface of the block walls by wetting them down lightly and evenly with a fine spray of water from a garden hose. Give the water a little time to soak in. If a section of the wall will not absorb water, rub it down with a wire brush. If it still will not absorb water, coat that entire wall with Weld-Crete or Permaweld-Z to ensure a bond between the stucco and the wall.

Stucco is prepared and applied in three layers. The first is called the scratch coat, the second the brown coat and the third the finish coat.

The first and second coats can be prepared simply from ordinary sand, provided it is clean, and mortar cement. However, a better bond between the stucco and the wall and thereby longer life will be secured by using stucco sand. Not only are the grains smaller but there is a greater variety of grain sizes in the sand sieved and selected for stucco than in standard sand. As a result the mixture is denser and stronger.

The top or finish coat can also be prepared from the same materials. However, a more attractive appearance will result from using white cement and white stucco sand and adding no more than 10 percent lime to the mixture. This produces an

Using a scarifier to produce deep horizontal lines in the first coat of mortar applied to a wall in the process of applying a stucco finish.

offwhite finish. To secure a colored finish add any of the mineral-oxide pigments available. Do not use any other type of pigment. Only the oxide pigment colors are stable over a long period of time. Keep a careful record of the quantity of color you add and mix very thoroughly, preferably by machine, to ensure an even finish color. As an alternative you can purchase ready-colored finish mix in dry form.

The scratch coat is made by mixing 1 part cement with 3 to 5 parts sand and adding sufficient water to make the mix workable, but little more. Start at the base of a wall and work upwards using a steel trowel and firm pressure. Make the scratch coat ½ inch thick. If the stucco will not adhere properly, reduce the sand in the mixture until it does.

Do one complete wall at a time. When you come to a projection, such as a window sill, bring the stucco as close as you can. Do not try for a tight joint, but after the job is completed seal the stucco to the sill with caulking compound. To make a straight corner, have a helper hold a long stick against the side of the building as a guide.

As soon as an area is completely covered it is scored with a scarifier to produce deep horizontal lines. The edge of a piece of metal lath or a piece of wood into which a number of nails have been driven can be used just as well.

The second or brown coat consists of the same mixture used for the scratch coat. It is applied with a float no less than four or five hours after the scratch coat has been applied. If you have to wait overnight or longer, moisten the surface of the scratch coat evenly with a fine spray of water, but do not soak it.

Again, you start at the bottom of the wall and work upwards, pressing the mix firmly against the wall. The brown coat is applied ⅜ inch thick and as evenly as possible. It should be applied to one entire wall at a time, otherwise a color difference may be seen in the finish coat at the point where you stopped and started again. Wait half a day or so and then carefully moisten the brown coat with a fine spray of water. Repeat this every few hours to keep the brown coat moist for two days. Then wait five days or more, moisten the brown coat evenly and proceed to apply the finish coat.

The finish coat need not be more than ⅛ inch thick. Again, you do an entire wall at one time or the results will be uneven. Start at the bottom and apply the finish mix with a steel trowel or a float. You can apply the mix evenly, or texture it in any number of ways. For example you can apply an even layer of stucco, fling small gobs of stucco against this layer and then use a float to bring the gobs almost flush with the rest of the surface. You can start by applying an uneven layer and produce uneven swirls by swinging the float in a series of circles. Lines can be made with a broom.

A concrete block wall can be given a simulated rough stucco finish by dipping a masonry brush into wet mortar and throwing the mortar onto the wall.

This pattern was produced by running a wood float in a series of circles over wet mortar splattered on a block wall.

Small colored stones can be thrown against the still wet stucco to produce a pebble effect. The wet stucco can be struck with small branch so that the leaves form imprints in the surface. This can be followed with light floating. Old-time masons very often leave their signature on a stucco job by finishing it off in a specific way.

Complete the curing by keeping the stuccoed walls moist for another five days. Doing so will greatly strengthen the stucco and reduce the possibility of it cracking.

ORNAMENTAL
CONCRETE
BLOCK WALLS

In a world where increasing property ownership leads to increasing and time-consuming upkeep work, ornamental walls of concrete block stand almost alone. Unlike a wood fence, which must be painted every few years, repaired just as often and replaced after an indecently short period of time, concrete-block walls require no attention, no repairs, no paint, and no replacement. They just stand there, indifferent to the weather, resistant to heat, cold, microbial attack, fire, little boys, and even fair-sized automobiles.

The major objection to concrete-block walls to date, and probably the major deterrent to their installation, despite their comparatively low cost and ease of construction, is appearance. These walls are generally not believed to be attractive. Perhaps they weren't in the past, but today, with the large selection of block designed especially for ornamentation, concrete-block walls can be highly interesting and attractive as the accompanying photos can testify.

DESIGN

Footings. All concrete-block walls must be supported by a footing sufficiently thick and broad to carry the load. The footing must be on firm subsoil, below the frost line. In a way, footings for ornamental walls are even more critical than build-

An attractive ornamental block wall made of pierced block capped by 1-inch solid block. *Courtesy National Concrete Masonry Assoc.*

ing footings. To some extent the very weight of the building on the foundation walls helps the footing resist frost pressure. Moreover whereas a foundation wall is always a part of a structure with four or more sides, the ornamental wall is a single entity. A crack in a foundation wall hardly reduces the overall strength of the foundation, witness the millions of homes with cracked foundations. A crack in an ornamental wall is not only unsightly, it seriously weakens the structure. Therefore do not stint on the digging nor on the size of the wall's footing.

376

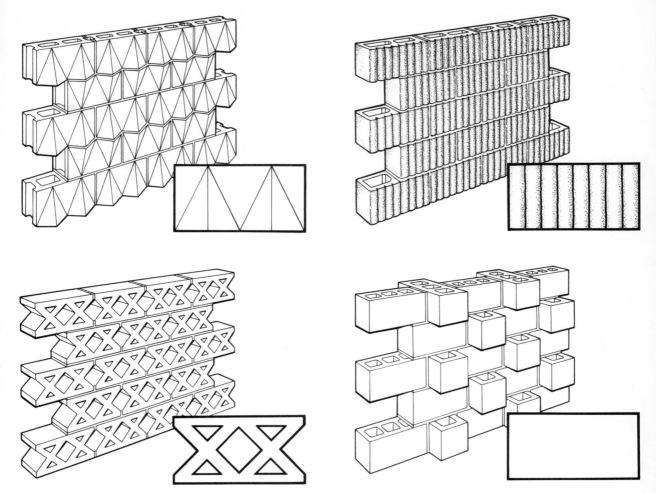

Patterned concrete blocks are available in an exciting variety of designs. These blocks can be used to build walls that screen while permitting the passage of light and air. The play of sunlight on the designs produces an attractive pattern of light and shadow, and your neighbor benefits from the beauty of the wall as much as you do.

The accompanying table suggests footing dimensions that will provide safe and stable supports for block walls up to 6 feet in height above grade. Should you want to build higher, it is best to secure local engineering aid. Note also that many municipalities control the construction of ornamental or freestanding — as they are sometimes called — walls higher than 5 feet. So, secure design approval before you start construction.

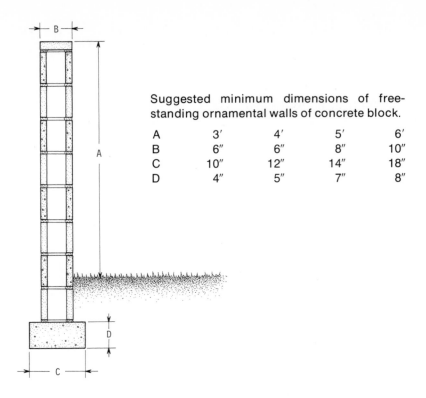

Suggested minimum dimensions of free-standing ornamental walls of concrete block.

A	3′	4′	5′	6′
B	6″	6″	8″	10″
C	10″	12″	14″	18″
D	4″	5″	7″	8″

The table assumes that the ground you uncover will be reasonably firm and dry. If it is not, provide the footing with a crushed stone base at least 4 inches thick and wider than the width of the footing by 6 or more inches.

The table also assumes that the bottom of the footing will be below the frost line. This is very important as a bad frost can easily crack and so destroy your wall. If you want to save a little on block, fill the bottom few inches of the trench with crushed stone or gravel, then pour on top of the stone.

Use a 1 : 3 : 5 mix a little on the wet side. Form construction, leveling, pouring, and screeding are done exactly as already suggested for other footings. If you use a trench footing without a form, pour carefully and make the mix wet enough to find its own level.

Choice of block. There are probably several hundred different ornamental block shapes currently manufactured. Unfortunately, every block plant does not make all types. Thus your economical choice of block is limited to what is locally available, and this selection may be further confined by price. Some blocks cost more than others. However, do not decide on the basis of unit cost alone. Compare the blocks on the basis of wall area per dollar. For example, a $1.00 block may cover only 18 by 8 inches, which works out to $1.00 per square foot of wall. A $2.00 block may cover 24 by 18 inches, which equals 3 square feet and works out to $.67 per square foot. And in making your final choice, remember that there is no point in saving money if you are not going to be pleased with the result.

Another factor to consider is block size. The larger the individual blocks the less work is involved.

Wall thickness. The height of the wall above grade more or less determines minimum wall thickness. (The wall below grade is supported laterally by the soil.) By rule of thumb, a 5-foot-high wall should be at least 8 inches thick to ensure stability. The wall doesn't have to be solid block, of course. A 6-foot-high wall should be 10 inches thick. Since the thickness of the block has a considerable bearing on its price, many stratagems have been devised to make thin walls strong enough to be self-supporting.

Reinforcing thin walls. One method is to incorporate block pilasters in the wall which acts as posts, strengthening the wall at that point. These posts are set every 6 or 8 feet depending on the size of the post and the thickness of the wall.

Another method, which can be used only with hollow walls, consists of reinforcing the wall with vertical steel bars every 6 or 8 feet. Generally, ½-inch-thick bars are used in walls up to 5 feet in height, and ¾-inch bars in walls up to 6 feet in height. To be effective the bars must penetrate the footing for at least 6 inches and run upwards through the wall to the next-to-last course. To secure the holes, pegs are placed in the still wet concrete and then removed as the concrete sets up. Then the block is laid as usual, care being taken to keep the holes open. When the wall is up except for the last course, the steel bars are slid down through the wall and into the holes. Then very wet mortar is poured between the steel and the footing and the block. This makes for a solid concrete wall at that point, reinforced by a steel bar locked into the footing.

A third method consists of curving the wall so that no portion is perfectly straight. The result is a beautiful wall, even when made from ordinary block, but it requires an equally curved footing and a lot of care in laying the block. The curved footing does not have to match the curve of the wall exactly, but the walls should never stray closer than 2 inches to the edge of the footing. The wall is made by first cutting a giant, curved template out of plywood to serve as a guide. This is supported

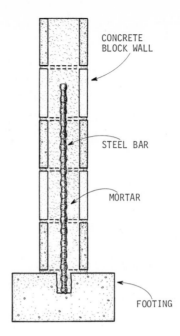

CONCRETE BLOCK WALL

STEEL BAR

MORTAR

FOOTING

When using steel bars to reinforce a block wall, leave holes or a channel in the footing to accommodate the bars and the mortar that will secure them.

by stakes driven into the ground. After each course is laid, the template is raised and then serves to guide the next course of block.

When planning a wall of this design, make certain the curve is not so sharp that the joints between the blocks will be unsightly.

Open versus closed patterns. This, in a way, is part of block selection. If you want a solid wall, you naturally select solid block. If you want your wall partially open to let in the air, pussy cats, and the stares of neighbors, you have a choice. You can either pattern solid blocks so that there are openings between them or you can purchase perforated blocks which usually look much better. In addition, perforated blocks tend to hide masonry errors, whereas standard blocks spaced apart tend to exaggerate misalignment and uneven spacing.

The choice of patterns can help or hinder your work. Stacked patterns, where all the blocks are placed vertically above one another, look very beautiful when all the joints are exactly the same width and depth and all the blocks are in perfect alignment. Perforated block can be stacked without any problem (some blocks are made to be stacked) because the block design takes the eye away from the joints, and these are so far apart that no one notices small variations. Stacking standard block, on the

380

other hand, is the kind of work even experienced masons dread. It is difficult and time-consuming, and the results are unsightly when the work is not perfect.

Making the joints a different color from the block is another practice to avoid, since it makes every joint error immediately obvious. On the other hand there is no reason to avoid the use of colored block. It is easy enough to match mortar color to block color. Use white cement, white sand and add the necessary color until you secure a close match. If you make the mortar color a bit lighter than that of the block, not only will you save on expensive color, the block will overshadow, and so detract attention from, the joints.

Closed pattern wall made of slumped concrete block which resembles adobe brick. *Courtesy National Concrete Masonry Assoc.*

An open pattern wall made of "split" block, a solid block that is actually broken lengthwise to produce the texture shown. Note the use of common brick in the wall to vary the pattern. *Courtesy National Concrete Masonry Assoc.*

Two more open, or perforated, block walls. The blocks are stacked, but this is not immediately apparent. *Courtesy National Concrete Masonry Assoc.*

When you choose to make your wall solid, you can change its appearance fairly easily by parging (plastering) its surface. In this case you can easily color the wall inexpensively and at the same time hide any patching you may have had to do. Follow the instructions given for stucco near the end of Chapter 25. The top of the wall can be gently rounded with the same colored mortar.

CONSTRUCTION TIPS

If you are building a solid wall and are going to plaster it when finished, there is no problem. You can start at the ends as usual and work towards a closure block. The plaster will hide everything. If you are constructing a wall of perforated or ornamental block which will not look good with a cut-down closure block in the middle, you have a choice of two methods to avoid this eyesore.

The first is to allow ½-inch-wide joints between the blocks, make the wall's length an exact multiple of the block and one side joint and then work very carefully. The other is to leave yourself plenty of room on the footing by making it one block length longer than necessary. Then start laying block at one end of the footing the usual way, but use a dummy wall end at the other end. The dummy end consists of a stake, driven vertically into the ground, with one side in line with the front of the wall. A nail is driven into the stake at the exact height of the first course. The guide line for the first course is stretched from the first block to the nail. Then the blocks are laid to this line. When you reach the end of the wall, you remove the stake. In this way, you can be off an inch or two on the first course and still not have to cut a closure block. The second and following courses are easily spaced correctly because you can follow the first course and see as each block is laid whether your side joints are too wide or narrow.

CAPPING THE WALL

You can lay slate or flagstone on top of the wall to serve as coping. You can simply mortar the top of the wall smooth. You can curve it using mortar and a trowel. You can cast an overhanging concrete cap on the wall by constructing a two-piece form around its top. The first part of the form is positioned with its top edge flush

TOPPING A WALL
WITH MORTAR

Spread a layer of fairly stiff mortar on the top of the wall.

Smooth and curve the mortar with a steel trowel.

The finished cap, made without any guides at all.

To cast an overhanging concrete cap atop a wall, construct a form as shown.

The finished cap, screeded and floated, with the form removed.

with that of the block. The second part of the form is nailed along the outside of the first. The top edge of the second part establishes the thickness of the cap. The thickness of the wood used for the first part establishes the distance the cap overhangs the wall. The form is then filled with a 1 : 2¾ : 3 mix, using ½-inch or smaller stones, screeded flat and floated smooth.

WORKING WITH BRICK

Man made and used bricks as early as 6000 B.C. The ancient city of Ur, legendary home of Abraham, was constructed almost entirely of brick. The biblical tower of Babel was probably built of brick. At least the ziggurats were. These were pyramidal structures often several hundred feet high, with astrological observatories on top that served some religious function during temple ceremonies. The ziggurats were constructed around a central core of sun-baked bricks and faced with highly colored enameled bricks. Their huge remains are still to be seen in the Tigris-Euphrates basin. The very largest may possibly have been the tower of Babel.

By 560 B.C., hard-burned bricks were common on the Mesopotamian plains. Babylon, under the great king Nebuchadnezzar, was renowned through the civilized world of that time for its beautiful and immense walls and temples, all of which were constructed of bricks. So highly was the art of brickmaking prized that kings of the period established royal brickworks and had their names stamped on the bricks.

Bricks haven't developed greatly since those near-forgotten times. We still manufacture them in much the same way from the same materials. Clay and shale, which is compressed clay that is first softened by exposure to the weather, are mixed together and molded into the desired shape. The brick is dried and then fired, which can take as much as a hundred hours. The technique has been refined and mechanized, but the method remains basically unchanged from its original conception.

The art of laying brick hasn't changed either, though the mortar has been considerably improved. Bricks are still laid one by one by hand—fortunately, because bricklaying is a very pleasant occupation. Professionals laying brick while perched on a shaking scaffold in a bitter winter's wind may not agree, but with your feet on the ground and the air filled with sunshine, laying brick can be extremely pleasant and relaxing—so much so that many people lay brick as a hobby. The late Winston Churchill was famous for his skill at bricklaying and held an honorary card in a British bricklayers' union.

If you are wondering what the bricklaying hobbyist does after he has completed all possible brick projects in and around his home, wonder no longer. A neighbor of mine constructed a 6-foot-high brick wall around his property, a brick patio, brick garden walls, brick steps, bricked his two-car driveway and erected two brick pillars alongside the entrance to his driveway. When all practical and useful brick projects were completed—he moved away. No doubt he is doing the same thing in another location.

Bricklaying is far easier than it may appear to be. True, speed and accuracy both take some time to acquire. They result from a trained combination of hand and eye that will give you considerable pleasure when you master it. But there is no need for speed when you are working for yourself. And accuracy isn't really necessary. A placed brick has to be very much out of line to catch one's eye. Remember, if the brick looks reasonably true to you when you lay it, it is going to look even better when the wall is completed. The very pattern of the bricks interlaced with the rectangles of mortar acts to hide discrepancies. The eye is deceived.

Of course you can accentuate errors in placement by using the wrong pattern. If you attempt a stacked pattern—one brick placed directly above another—you are looking for visual trouble. The work has to be perfect to look good. But all the other patterns work for you, so there is really no need for stacking brick.

As with concrete block, accuracy of placement has very little to do with the strength of the finished job. The wall you construct will be just as strong if the joints vary visibly and the bricks are not perfectly aligned.

It is physically easier to lay brick than it is to lay block. Bricks are, of course, considerably lighter. It is no effort at all to place one carefully in position. This isn't so with block. It takes two hands to position even the smaller blocks, and 10-inch and larger blocks require two men for controlled placement.

So all in all, you will find laying brick a pleasant and satisfying experience and far less difficult than you may have imagined it to be.

Most of the tools used for laying block are used for laying brick. They include:

Mortar pan   Joint tool
Brick hammer   Line level
Brick trowel   Mason's brush
Pointing trowel

LAYING BRICK

BRICK TYPES AND SIZES

Although there are reportedly over 10,000 different types of bricks in various sizes currently manufactured here and abroad, they can all be divided into four basic categories: common or building brick, face brick, fire brick, and paving brick.

Building brick can be and is used for everything. It is strong and durable but tends to be somewhat less than perfectly straight and to vary slightly in size, shape, and even color from brick to brick. Sometimes the bricks are also a little warped, chipped, and cracked. Dimensional differences between one brick and the next can be as much as one half inch. But building brick is the least expensive brick of all and is therefore used far more often than any other category.

Whereas building bricks are never anything more than clay molded to a rectangular shape, face bricks, which are used for exterior surfaces, can be works of art. There is no limit to the variety of textures and colors used with face brick. The body of the brick itself may range from pale yellow to almost black or purple. Its surface may be smooth or textured in any of a hundred different patterns, and one or more sides may be glazed in colors ranging from china white to midnight black.

Face brick is more carefully controlled during manufacture, and therefore its dimensions, texture, and color are uniform from brick to brick and from pallet load to pallet load. Face bricks are the most expensive type of bricks made. However, they are not much stronger than building bricks, though they resist weathering far better than the best of the building bricks.

Fire bricks are pale yellow bricks made from special clays carefully fired. They are used to line furnaces, ovens, and fireplaces. They are specifically designed to resist heat. However, to preserve this ability they must be laid up with either fire clay or a special aluminum oxide mixture made for the purpose. Ordinary mortar will disintegrate quickly under heat. For ordinary work, such as fireplaces and barbecues, the half-thick bricks can be used.

Paving brick is an exceptionally strong brick designed especially for paving roads. It is mentioned here just for the record. There is no need for it should you decide to pave your drive. It is necessary only for commercial roadways.

Building bricks are manufactured in three grades: SW (severe weathering), MW (moderate weathering), and NW (no weathering). Face bricks can be had in two grades, and fire brick is made in one grade only.

Where the building code permits and where there is little frost, you can safely use the cheapest brick available for all your work. Where the frost is severe it is necessary to use SW brick outdoors, but NW brick can be used indoors and as the inner courses of a multi-brick exterior wall. NW brick can be used outdoors in frost country only if the brick is protected from rain and snow. It is not the frost that breaks the bricks down, it is exposure to water followed by hard frost that causes them to slowly disintegrate. Bear in mind, however, that even the hardest brick should be capped by a watertight covering of some kind and should not be exposed to direct rainfall.

While brick types are reasonably straightforward, brick size can be confusing. This results from the practice of assigning two or more sets of dimensions to a single brick. The brick may be defined by either its actual size or its nominal size. The actual size is the number you would read if you placed a ruler to it. The nominal size includes the thickness of the mortar joint and is sometimes given as brick size within the wall. Since joint thickness can be specified as $1/4$, $3/8$, or $1/2$ inch, just being told the nominal size of the brick doesn't help too much.

The best way to find the actual size of a brick is to ask your supplier and work from the dimensions he provides. It is useless to measure a single building brick yourself since it can be close to $1/4$ inch away from its specified dimensions. For example, the size of modular bricks is $2^{1/4}$ by $3^{5/8}$ by $7^{5/8}$ inches. This means that the average length of bricks in a pallet load will measure out to $7^{5/8}$ inches. However, there will be some 8-inch-long bricks and some $7^{1/2}$-inch bricks in the pile. It can therefore be misleading to measure just one brick.

Selecting the brick. There are only two reasons to use anything but building brick. One is appearance and the other is economy. If you want brick that looks different from common brick and you do not want to build with used bricks—which is the cheapest of all brick—you have no choice but to pay the extra money and purchase face brick. As for economy, much depends on the construction you have in mind. In certain circumstances, the larger and individually more costly brick can prove more economical. For example, the SCR brick, which is 5½ inches wide without mortar, can in many localities be used to build a single-brick-wide wall for a single-story home. In such cases the SCR brick replaces two 3¼-inch-wide building bricks.

FOOTING REQUIREMENTS

Like concrete block, mortared bricks cannot be laid directly on the ground but require suitable footing.

Load-bearing brick walls which can be and are used for building foundations, building walls, and similar structures are always supported by footings fairly close in dimensions to those used for block walls. As a rule of thumb, the thickness of the footing should be equal to that of the wall and its width should be twice that dimension. A three-brick, load-bearing wall with a nominal thickness of 12 inches could be safely carried by a footing 24 inches wide and 12 inches thick. Where the code permits and the earth can carry the load, footing dimensions can be somewhat reduced. But if you aren't certain, you can be sure this rule will give you plenty of support.

Non-load-bearing walls such as garden walls and room dividers can be safely carried by footings similar to those suggested for ornamental concrete-block walls in Chapter 26.

Use the concrete mixtures already suggested for block footings.

Standard	2¼"x3¾"x8"
Modular	2¼"x3⅝"x7⅝"
Jumbo	2¾"x3¾"x8"
Norman	2¼"x3⅝"x11⅝"
SCR	2⅛"x5½"x11½"
Roman	1⅝"x3⅝"x11⅝"
Baby Roman .	1⅝"x3⅝"x7⅝"
Fire Brick	2½"x3⅝"x9"
Oversize	Sizes vary with manufacturer

Actual sizes of the more frequently used types of bricks. Standard brick is often called common.

MORTAR

Use exactly the same cement-lime-sand mixture as that used for laying block. To save you the trouble of referring back to Chapter 23, the mix is repeated here. For the strongest joints possible use Type M mortar, which consists of 1 part cement, ¼ part hydrated lime, and 3 parts sand. Or use mortar cement, which has the necessary lime mixed in.

PREPARING THE BRICK

To secure maximum adhesion between brick and mortar it is necessary to moisten the brick. (This is not necessary with block.) At the same time, you don't want the brick too wet. A simple check consists of placing a few drops of water on the brick. If the drops disappear in less than a minute the brick is too dry for best results and must be wetted down. Do this by turning the garden hose on the bricks. Let the water pour over them and do not stop hosing until you can see water running off them, indicating that they can absorb no more water. Then, by the time you get yourself up and ready to lay brick, the bricks will have lost sufficient moisture to be ready for the mortar.

LAYING BRICK

The accompanying photos show the construction of a short, low, one-brick-thick wall. The same basic procedures and techniques are used in all brick work.

Start by marking on the footing the front or rear corners of the wall that is to be erected. Drive nails into the footing if you can. If not, make a small hole or scratch mark. Snap a chalk line between the two marks or use a board as a guide and scratch a line between the two marks.

Take a number of bricks that you are going to use and lay them end to end between the nails or scratches indicating the ends of the wall. Space the bricks evenly apart. If there isn't at least ⅜ inch of space between the ends you either have to select shorter bricks or extend the length of the wall. If you don't do either, you will end up using less than a full brick or an exact half brick in the wall. Should you opt to extend the wall (or shorten it—your choice), bear in mind that it is difficult to work with joints narrower than ½ inch.

HOW TO LAY BRICK

1 Mark the terminal points of the wall with a couple of nails driven into the footing and, using a long straight-edge, scratch a line between the two points. This line marks one edge of the wall to be erected.

2 Starting at one terminal point, spread an inch-thick bed of mortar the length of a brick and a half, along the guideline. See drawings on next page for correct method of "throwing" mortar onto a brick.

HOW TO "THROW" MORTAR

Slice a gob of mortar from the board or box.

Turn the trowel edgewise and, with a short sweep . . .

. . . throw the mortar onto the brick.

Using the flat of the trowel, spread the mortar.

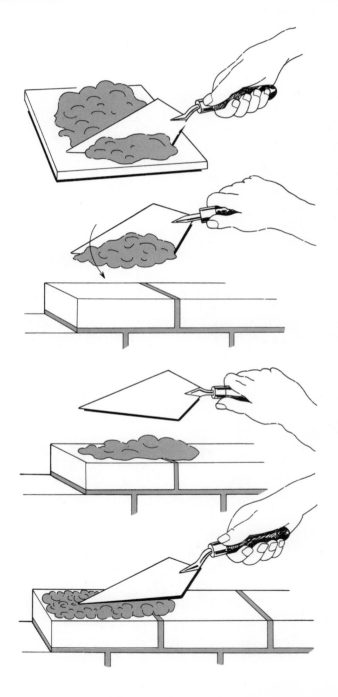

3 Lay the first brick onto the bed of mortar, pressing it down until it is about a half inch above the footing. Level it in two directions, tapping a high corner with the trowel handle.

4 Lay a second brick at the other terminal point; level it. Then stretch a guideline between the two bricks, using two other bricks as shown. The line should barely touch the in-place bricks. Align the in-bricks with the line.

5 Spread an inch-thick bed of mortar on the footing next to one of the in-place bricks. Then butter the end of a brick with mortar, pressing it down with the trowel so it adheres.

6 Place the brick on the mortar, buttered end about 2 inches from the in-place brick. Press the brick lightly into the mortar and slide it gently toward the in-place brick, stopping a joint thickness (½ inch) from the end. If you haven't buttered the brick sufficiently, and the joint is too small, start again. Level the brick in two directions.

7 Continue laying bricks until you are one short of the end brick. Now you must lay the closure brick. Check for fit and cut the closure brick, if necessary. Spread a bed of mortar; butter one end of the closure brick, the opposite end of one of the adjacent bricks, and lower the closure brick into place. Align and level the closure brick.

8 The second course is usually started with a half brick, in order to stagger the joints and form what is known as a running bond. Cut the brick with a brick set as shown here, or with a mason's hammer.

9 Scoop and throw a long gob of mortar on the end brick and place the half brick in position, tapping it into place to get a half-inch joint. Then level the half brick in two directions. Repeat the operation with another half brick at the other end of the wall.

10 Run a line between the two half-bricks, using two bricks as you did when laying the first course. This line will serve as the guide for laying the second course.

11 When you have laid the second course, build up the corners of the wall. Use the level to check the bricks vertically and horizontally.

12 As you work, clean the sides of the wall with the edge of the trowel, scraping the joints flush with the wall. Don't let mortar drip from the joints; it will stain the bricks.

13 Tool the joints shortly after you scrape them. Bricks absorb water quickly; if you wait too long, tooling will be difficult. After tooling, allow the joints to dry and then brush the wall to get rid of excess mortar.

14 Continue laying bricks in this manner, cleaning as you go, until you have reached the desired height. Insert the last closure brick and scrape, tool, and brush the wall.

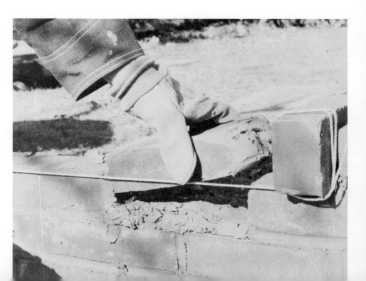

Cleaning. You can reduce to a minimum the need to clean brickwork by cleaning, tooling, and brushing the bricks as soon as possible after laying them. However, there will be times when it will be necessary to remove mortar stains from the bricks no matter how careful you may be.

To do so, try strong soap and hot water applied with a scrub brush first. If the mortar hasn't become too hard, this will work. If it doesn't, use one of the commercial brick-cleaning fluids sold at the mason's supply yard or a solution consisting of 1 part muriatic acid and 9 parts water. Always add the acid to the water and always use rubber gloves. Apply carefully with a stiff brush and wash clean afterwards.

Finishing the joints. Joints are finished or tooled, as the process is also called, for two reasons: appearance and strength. Merely striking, which means scraping the sides of the bricks clean of mortar with the edge of the trowel, leaves a smooth, flush joint surface, but it doesn't strengthen the joint. Tooling, which consists of pressing the mortar into the space between the bricks, forces the mortar tightly between the bricks and makes for a much stronger, denser joint.

However, while all tooling compacts the joint, some tooled joints have less resistance to the weather than others because of their shape. Concave and V-joints have the greatest resistance to rain penetration. Where maximum weather resistance is not important any of the tooled joint shapes can be used. However, this writer strongly advises against extruded joints. They cannot be cleaned without a tremendous amount of labor.

Using a story pole. A story pole is simply a vertical pole on which the course heights of the brick wall under construction are marked. When you are building a

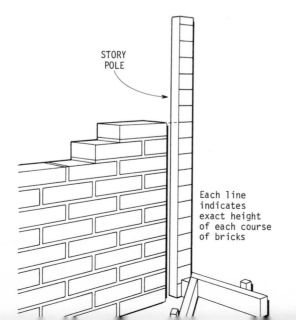

STORY
POLE

Each line
indicates
exact height
of each course
of bricks

A story pole enables you to control the height of a brick wall by checking the course markers and varying joint thickness.

401

garden wall, for example, there is no need to control the height of the wall. It matters little if the wall is a few inches higher or lower. But when you must terminate a wall at a specific height, it is important to keep track of your course elevations as you go.

Let us suppose you want to hit 11 feet on the nose. This requires exactly forty-eight courses of 2¼-inch brick laid with ½-inch-thick joints. You can keep measuring as you go or you can mark off the forty-eight courses on a long, clean, straight stick and position the stick vertically alongside your growing wall. Then, as you work, you keep looking at the course markers. In this way you can quickly see whether or not your joint thicknesses are exactly right, a bit thin, or a bit thick. If marking off forty-eight courses on a stick sounds like too much work, you can purchase premarked metal tape that you can apply to a suitable stick.

BRICK
WALLS

The preceding chapter covered the technique of laying brick. This chapter covers the application of that technique to building walls. An outdoor cooking center or barbecue consists of four short walls joined to form a square or rectangle. A home storage shed, pool-side cabana, or dog house consists of four longer walls. The list of things you can make using walls is almost endless.

Wall area sq. ft.	Wall thickness							
	1 brick		2 bricks		3 bricks		4 bricks	
	Number of bricks	Mortar cu. ft.	Number of bricks	Mortar cu. ft.	Number of bricks	Mortar cu. ft.	Number of bricks	Mortar cu. ft.
1	6.17	.08	12.33	.2	18.49	.32	24.65	.44
10	61.7	.8	123.3	2	184.9	3.2	246.5	4.4
100	617	8	1,233	20	1,849	32	2,465	44
200	1,234	16	2,466	40	3,698	64	4,930	88
300	1,851	24	3,699	60	5,547	96	7,395	132
400	2,468	32	4,932	80	7,396	128	9,860	176
500	3,085	40	6,165	100	9,245	160	12,325	220
600	3,712	48	7,398	120	11,094	192	14,790	264
700	4,319	56	8,631	140	12,943	224	17,253	308
800	4,936	64	9,864	160	14,792	256	19,720	352
900	5,553	72	10,970	180	16,641	288	22,185	396
1,000	6,170	80	12,330	200	18,490	320	24,650	440

Brick and mortar quantities required for walls made of standard brick (2¼ by 3⅝ by 8 inches) using ½-inch-thick joints. When using ⅝-inch joints, multiply quantity by 80%. When using ⅜-inch joints, multiply quantity by 120%. Note, mortar figures include only approximately 10% additional mortar for waste.

When the size of the bricks you plan to use differs from the brick size given in the table and you want to know fairly accurately how many bricks will be needed to fill a given area, one brick thick, draw a picture of the brick and add the mortar thickness you plan to use on two sides. Convert the overall dimensions to decimals for easy computation. Multiply length by height to get square inches. Then divide 144 by the number of square inches per brick and mortar to determine the number of bricks per square foot.

For example, with mortar on top and the side the dimensions of the Norman brick shown are 12.125 by 2.75 inches, which works out to 33.343 square inches. Dividing 144 by this number produces 4.31, which is the number of these bricks needed to cover a square foot when set in ½-inch-thick mortar.

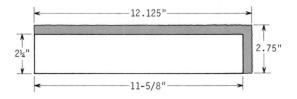

Courses	Height	Courses	Height	Courses	Height	Courses	Height	Courses	Height
1	0'2⅝"	21	4'7⅛"	41	8'11⅝"	61	13'4⅛"	81	17'8⅝"
2	0'5¼"	22	4'9¾"	42	9'2¼"	62	13'6¾"	82	17'11¼"
3	0'7⅞"	23	5'0⅜"	43	9'4⅞"	63	13'9⅜"	83	18'1⅞"
4	0'10½"	24	5'3"	44	9'7½"	64	14'0"	84	18'4½"
5	1'1⅛"	25	5'5⅝"	45	9'10⅛"	65	14'2⅝"	85	18'7⅛"
6	1'3¾"	26	5'8¼"	46	10'0¾"	66	14'5¼"	86	18'9¾"
7	1'6⅜"	27	5'10⅞"	47	10'3⅜"	67	14'7⅞"	87	19'0⅜"
8	1'9"	28	6'1½"	48	10'6"	68	14'10½"	88	19'3"
9	1'11⅝"	29	6'4⅛"	49	10'8⅝"	69	15'1⅛"	89	19'5⅝"
10	2'2¼"	30	6'6¾"	50	10'11¼"	70	15'3¾"	90	19'8¼"
11	2'4⅞"	31	6'9⅜"	51	11'1⅞"	71	15'6⅜"	91	19'10⅞"
12	2'7½"	32	7'0"	52	11'4½"	72	15'9"	92	20'1½"
13	2'10⅛"	33	7'2⅝"	53	11'7⅛"	73	15'11⅝"	93	20'4⅛"
14	3'0¾"	34	7'5¼"	54	11'9¾"	74	16'2¼"	94	20'6¾"
15	3'3⅜"	35	7'7⅞"	55	12'0⅜"	75	16'4⅞"	95	20'9⅜"
16	3'6"	36	7'10½"	56	12'3"	76	16'7½"	96	21'0"
17	3'8⅝"	37	8'1⅛"	57	12'5⅝"	77	16'10⅛"	97	21'2⅝"
18	3'11¼"	38	8'3¾"	58	12'8¼"	78	17'0¾"	98	21'5¼"
19	4'1⅞"	39	8'6⅜"	59	12'10⅞"	79	17'3⅜"	99	21'7⅞"
20	4'4½"	40	8'9"	60	13'1½"	80	17'6"	100	21'10½"

This table shows the height of brick walls of 1 to 100 courses using 2¼-inch brick and ⅜-inch joints.

Courses	Height	Courses	Height	Courses	Height	Courses	Height	Courses	Height
1	0'2¾"	21	4'9¾"	41	9'4¾"	61	13'11¾"	81	18'6¾"
2	0'5½"	22	5'0½"	42	9'7½"	62	14'2½"	82	18'9½"
3	0'8¼"	23	5'3¼"	43	9'10¼"	63	14'5¼"	83	19'0¼"
4	0'11"	24	5'6"	44	10'1"	64	14'8"	84	19'3"
5	1'1¾"	25	5'8¾"	45	10'3¾"	65	14'10¾"	85	19'5¾"
6	1'4½"	26	5'11½"	46	10'6½"	66	15'1½"	86	19'8½"
7	1'7¼"	27	6'2¼"	47	10'9¼"	67	15'4¼"	87	19'11¼"
8	1'10"	28	6'5"	48	11'0"	68	15'7"	88	20'2"
9	2'0¾"	29	6'7¾"	49	11'2¾"	69	15'9¾"	89	20'4¾"
10	2'3½"	30	6'10½"	50	11'5½"	70	16'0½"	90	20'7½"
11	2'6¼"	31	7'1¼"	51	11'8¼"	71	16'3¼"	91	20'10¼"
12	2'9"	32	7'4"	52	11'11"	72	16'6"	92	21'1"
13	2'11¾"	33	7'6¾"	53	12'1¾"	73	16'8¾"	93	21'3¾"
14	3'2½"	34	7'9½"	54	12'4½"	74	16'11½"	94	21'6½"
15	3'5¼"	35	8'0¼"	55	12'7¼"	75	17'2¼"	95	21'9¼"
16	3'8"	36	8'3"	56	12'10"	76	17'5"	96	22'0"
17	3'10¾"	37	8'5¾"	57	13'0¾"	77	17'7¾"	97	22'2¾"
18	4'1½"	38	8'8½"	58	13'3½"	78	17'10½"	98	22'5½"
19	4'4¼"	39	8'11¼"	59	13'6¼"	79	18'1¼"	99	22'8¼"
20	4'7"	40	9'2"	60	13'9"	80	18'4"	100	22'11"

Wall height for 1 to 100 courses using 2¼-inch brick and ½-inch joints.

Courses	Height	Courses	Height	Courses	Height	Courses	Height	Courses	Height
1	0'2⅞"	21	5'0⅜"	41	9'9⅞"	61	14'7⅜"	81	19'4⅞"
2	0'5¾"	22	5'3¼"	42	10'0¾"	62	14'10¼"	82	19'7¾"
3	0'8⅝"	23	5'6⅛"	43	10'3⅝"	63	15'1⅛"	83	19'10⅝"
4	0'11½"	24	5'9"	44	10'6½"	64	15'4"	84	20'1½"
5	1'2⅜"	25	5'11⅞"	45	10'9⅜"	65	15'6⅞"	85	20'4⅜"
6	1'5¼"	26	6'2¾"	46	11'0¼"	66	15'9¾"	86	20'7¼"
7	1'8⅛"	27	6'5⅝"	47	11'3⅛"	67	16'0⅝"	87	20'10⅛"
8	1'11"	28	6'8½"	48	11'6"	68	16'3½"	88	21'1"
9	2'1⅞"	29	6'11⅜"	49	11'8⅞"	69	16'6⅜"	89	21'3⅞"
10	2'4¾"	30	7'2¼"	50	11'11¾"	70	16'9¼"	90	21'6¾"
11	2'7⅝"	31	7'5⅛"	51	12'2⅝"	71	17'0⅛"	91	21'9⅝"
12	2'10½"	32	7'8"	52	12'5½"	72	17'3"	92	22'0½"
13	3'1⅜"	33	7'10⅞"	53	12'8⅜"	73	17'5⅞"	93	22'3⅜"
14	3'4¼"	34	8'1¾"	54	12'11¼"	74	17'8¾"	94	22'6¼"
15	3'7⅛"	35	8'4⅝"	55	13'2⅛"	75	17'11⅝"	95	22'9⅛"
16	3'10"	36	8'7½"	56	13'5"	76	18'2½"	96	23'0"
17	4'0⅞"	37	8'10⅜"	57	13'7⅞"	77	18'5⅜"	97	23'2⅞"
18	4'3¾"	38	9'1¼"	58	13'10¾"	78	18'8¼"	98	23'5¾"
19	4'6⅝"	39	9'4⅛"	59	14'1⅝"	79	18'11⅛"	99	23'8⅝"
20	4'9½"	40	9'7"	60	14'4½"	80	19'2"	100	23'11½"

Wall height for 1 to 100 courses using 2¼-inch brick and ⅝-inch joints.

SINGLE-BRICK WALLS

The most famous single-brick-thick wall ever built is probably the one constructed by Thomas Jefferson on his Monticello estate. To conserve brick and still retain sufficient lateral stability for safety, Jefferson constructed his wall in a series of curves. Purportedly, it was the first of its kind ever constructed.

To build a similar, sinusoidal wall, start by making a full-size template of plywood. It should be at least two full curves long and strong enough to be moved about and easily supported above the earth without breaking. Use the template to guide you in laying out an 8-inch-wide, 6-inch-thick concrete footing. If the soil is sufficiently firm, you can simply dig the necessary snakelike trench and pour the footing into the bottom of it. If not, you can use a number of planks for the footing form. It isn't necessary that the footing be curved nor that the wall be centered perfectly on the footing. The wall can be off center a few inches without causing any problems. What is very important, however, is that the footing be positioned below the frost line.

Use the template in place of the stretched line as your guide while laying the brick. Hold the wall to a 4-foot height. If you want to go higher, it is best to secure local engineering advice.

The only way to construct a straight, single-brick wall over 3 feet in height and have it stand up to bicycles, children, and strong winds is to reinforce the wall every few feet with an integral pilaster. You need a few more bricks to make this kind of wall than a sinusoidal wall, but it is a lot easier to build.

Make the footing 8 inches wide and 6 inches thick beneath the wall itself and 16 inches square under the pilasters. The thickness here is kept the same. Position the pilasters 6 feet apart on 4-foot-high walls and 5 feet apart on a 5-foot-high wall. Don't forget to position the footing below the frost line.

Right-angle turns in a single-brick wall are made by overlapping the bricks at the corner in alternate courses, as shown in the drawing.

As an alternative, the corner can be positioned within a pilaster, which then acts as an ornamental corner post. The wall should terminate with a pilaster for better appearance and greater strength.

Screen-type single-brick walls are made by omitting bricks from the wall in a regular pattern. As can be imagined, these walls are not very strong, and if you want to go to 4 feet in height without curves and without pilasters it is best to use SCR bricks for the wall.

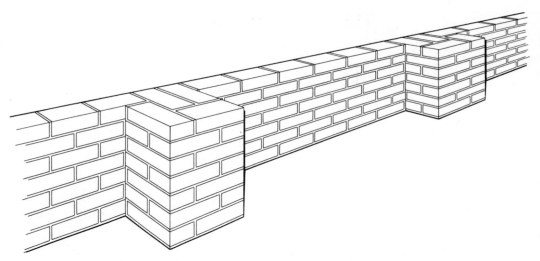

Pilasters must be built into a single-brick wall that is higher than 3 feet and that will be subjected to stress. The pilasters can be built in different ways, as shown in the drawing below.

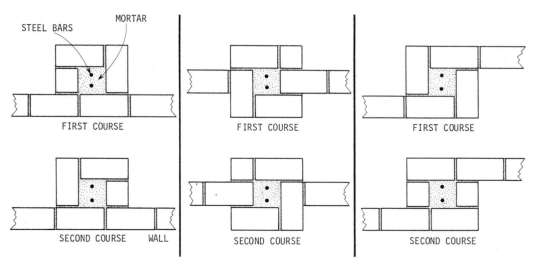

Three ways of introducing a pilaster into the middle of a one-brick wall. The drawings show the way the bricks are arranged in the first and second courses. This arrangement is then repeated in subsequent courses.

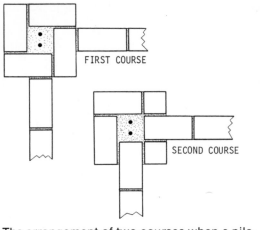

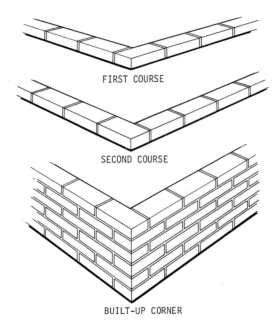

FIRST COURSE

SECOND COURSE

The arrangement of two courses when a pilaster is built into the corner of a wall.

Right: Steps in building up a corner in a one-brick wall.

FIRST COURSE

SECOND COURSE

BUILT-UP CORNER

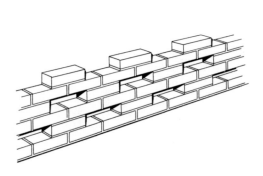
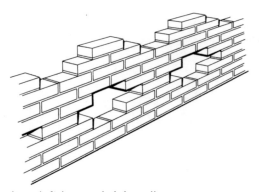

Two examples of patterned openings left in one-brick walls.

TWO-BRICK WALLS

Two-brick walls can be constructed in almost the same way as single-brick walls. If you are constructing a load-bearing wall, such as the wall of a building, make the footing 16 inches wide and 8 inches thick. If you are constructing a non-load-bearing wall, for example a garden wall, a 14-inch-wide, 6-inch-thick footing will be sufficiently strong and stable.

Start by snapping a line down the length of the footing. Off-center the line sufficiently to place the center of the two-brick wall in the center of the footing. Construct a single-brick wall three or four courses high, carefully cleaning both sides of the wall as you go. Wait a little while to let the mortar stiffen a bit. Then lay the second half of the wall. This is easier to do when you do not have to worry about touching the first half of the wall and possibly pushing the bricks out of place. If you have started the first wall with a whole brick, start the second with a half brick. You do not want the vertical joints of the two sides of the wall to line up.

Butter the ends and sides of the bricks forming the second wall so that there is mortar between the bricks forming the two sides of the wall. The separation between the two halves of the wall is critical only when you install headers. These are bricks placed across the width of the wall in bonding patterns that result in a stronger and possibly more attractive wall. The simplest, known as a common bond, consists of a row of headers laid in a solid row every fourth or fifth course. When the headers are alternated with stretchers (bricks laid lengthwise with the wall) the resultant pattern is called a Flemish bond. No matter which pattern you select, and four others are shown, you must make certain you space the two stretcher-brick walls correctly so that the ends of the headers lie flush with the sides of the stretchers.

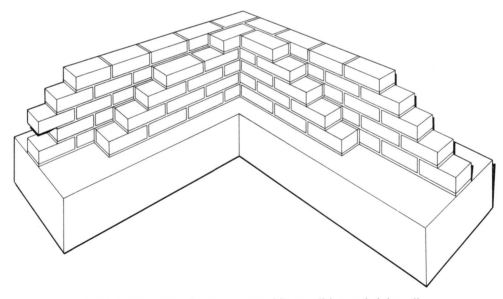

How a corner can be constructed in a solid, two-brick wall.

FIVE FREQUENTLY USED BONDS

Running bond. It consists of all "stretchers" (that is, brick laid lengthwise along the wall). Running bond is frequently used in veneered walls and in interior walls.

Common bond. A variation of running bond, it has a course of "headers" (brick laid with the short end along the face of the wall) at regular intervals. These header courses may appear every fifth, sixth, or seventh course.

Flemish bond. A handsome bond often used in Colonial American buildings, each course of which is made up of alternate stretchers and headers, with the headers in alternate courses centered over the stretchers above and below.

English bond. Composed of alternate courses of headers and stretchers, with headers centered on the stretchers and joints between stretchers in all courses line up vertically.

Stack bond. Created by using either all stretchers or all headers, and aligning all joints vertically. This bond is used only in veneered or other nonstructural walls.

Courtesy Brick Institute of America

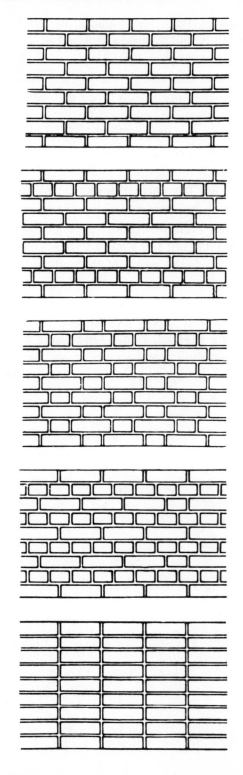

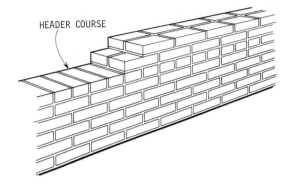

HEADER COURSE

How header courses are used to strengthen a two-brick, common-bond wall.

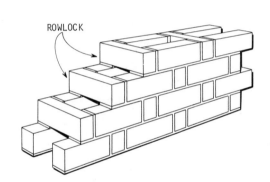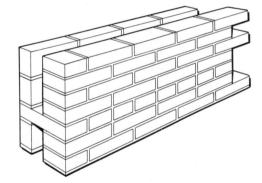

ROWLOCK

Two forms of rowlock walls.

The two-brick wall can be depended on to stand securely by itself to a height of 5 feet. For a stronger wall incorporate pilasters every 10 or 12 feet. Make the footing beneath the pilasters at least 4 inches wider in both directions than the base of the pilaster. With pilasters, a straight garden wall two-bricks thick should be safe to a height of 6 feet.

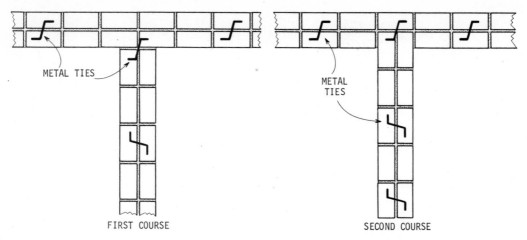

FIRST COURSE   SECOND COURSE

How intersections in a two-brick wall can be handled.

The rowlock or rolok-bak wall is a hollow wall consisting of two single-brick walls tied together by headers. In one version, the bricks of one wall are laid flat in the usual way. The bricks of the second wall are laid on edge. With this arrangement, the two walls are of equal height every seventh course, which is the header course. An alternative arrangement consists of placing the bricks of both walls on edge. The headers can then be placed at any desired course.

Cavity walls can be constructed with the bricks of the two walls laid on edge or flat. No headers are used because that would negate the main reason for cavity walls, which is insulation. Instead, the two walls are joined by metal ties specially made for the purpose. Use the ties every second or third course and every 2 feet of wall length.

THREE-BRICK WALLS

Three-brick walls are used for buildings of more than one story and for high garden walls. The construction procedure is essentially the same as for single- and dual-brick walls. However, it is easy to become confused as to which bricks must be cut and which brick goes where, especially at the corners, so it is advisable first to lay the bricks dry just to make certain everything will work out properly.

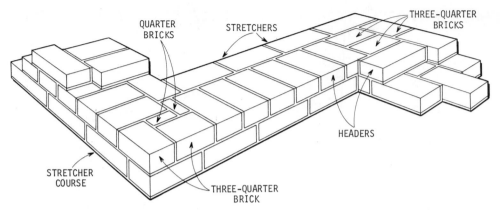

QUARTER BRICKS

STRETCHERS

THREE-QUARTER BRICKS

HEADERS

STRETCHER COURSE

THREE-QUARTER BRICK

How to lay brick to form a 3-brick-wide (12-inch) common-bond wall. The fractional bricks are used as closures and are placed in position last.

PIERS AND COLUMNS

The construction of brick piers and columns starts with a footing that is at least 6 inches larger all around than the base of the pier or column and thick enough to carry the load. To be certain of this, make the pier footing as thick as the footings used for the other portions of the building or half as thick as its width.

The bricks that form the pier can be arranged any way that suits your purpose as long as the joints are staggered. Make certain the structure is as vertical as you can possibly make it. To reinforce a pier, insert a number of steel reinforcing bars into the center of the footing while the concrete is still soft. Hold the rods in a vertical position until the concrete hardens. Then build the pier around the rods, filling the space between the steel and the brick with mortar.

CAPPING A WALL

The top surface of all exposed brick walls should be capped in order to prevent rain from soaking into the highly absorbent brick and mortar. Should a frost follow the rain the wall will be damaged.

THREE WAYS TO CAP AN EXPOSED BRICK WALL

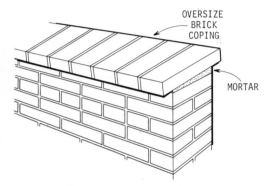

OVERSIZE
BRICK
COPING

MORTAR

Bricks set on edge at an angle above the
top of the wall.

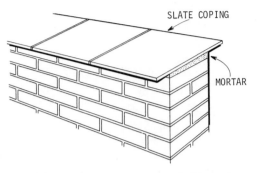

SLATE COPING

MORTAR

Slate mortared in place at a slight angle.

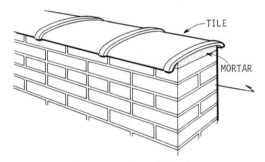

TILE

MORTAR

Overlapping tile mortared in place.

One method of capping consists of placing hard-fired brick at an angle across the top of the wall. Rain falling on the pitched surface runs quickly off and is not absorbed.

Another capping method consists of placing flagstone or slate atop the wall and holding it in place with mortar. The stone should be given a slight pitch. Still another consists of casting a concrete cap on top of the wall. The best and most expensive method consists of using glazed tile as a capping. Like roof shingles these tiles are placed so that the end of one overlaps another. They are held in place by mortar.

BRICK
VENEER

Brick veneer can serve many purposes. It can improve the appearance of an existing home. It can hide an ugly basement wall. It can decorate a living room or kitchen wall. It can be used in place of siding, in which case it eliminates almost all the painting and other maintenance necessary with clapboard and similar sidings. Brick veneer, of course, also adds to the fire resistance of the building.

REQUIREMENTS

Although brick veneer is no more than a brick wall a single brick thick and is constructed exactly the same way as any other brick wall, veneer construction must satisfy the following conditions if the wall is not to crack and the results are to be attractive:

1. Brick veneer must be carried by solid masonry of one kind or another.

2. The veneer must stand free and clear of all walls. The recommended space is 1 inch, but anything down to about ½ inch seems to work acceptably.

3. The veneer must be fastened to the adjacent wall by suitable metal ties at frequent intervals.

4. The veneer must come up behind the roof fascia.

Care must be exercised in deciding on joint thickness and in laying the brick to make certain that no patching will be necessary.

Indoors a brick veneer wall can be laid directly on a concrete cellar floor, on any portion of a slab foundation, and on a masonry hearth (to veneer a fireplace).

Outdoors veneer can be laid over existing siding, on any solid masonry porch or stoop. When there is no such support available, a solid support must be provided in some way.

A brick veneer wall that covers the lower part of a new house. Siding will be used to cover the upper walls.

On old work, excavate down to the footing and clean the ledge and adjoining foundation wall. Apply Weld-Crete or Permaweld-Z, which are bonding agents. Then construct a 6-inch block wall on the footing, making certain you bond the block to the wall with mortar as you go. Quit when you are flush or close to flush with the top of the foundation wall. You now have a base 6-inches wide on which to lay the brick. Assuming you will use 3¼-inch-wide brick, the 6-inch base will give you the clearance you need between the back of the brick and the siding.

As an alternative to digging all the way down to the footing, you can bolt an angle iron to the side of the foundation about 1 foot below grade. Lay the brick on the metal ledge. Then cover the metal with tar to protect it from corrosion.

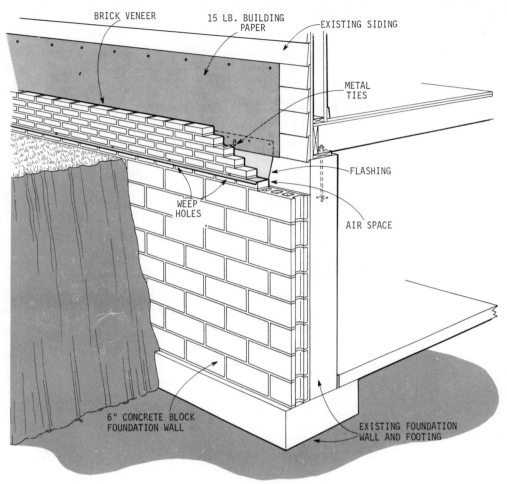

BRICK VENEER · 15 LB. BUILDING PAPER · EXISTING SIDING · METAL TIES · FLASHING · AIR SPACE · WEEP HOLES · 6" CONCRETE BLOCK FOUNDATION WALL · EXISTING FOUNDATION WALL AND FOOTING

To provide a support for brick veneer on an existing house, a foundation of 6-inch concrete block is built on the footing adjacent to the old foundation wall.

On new work, simply make the foundation 4 inches thicker to provide a 4-inch-wide ledge for the bricks. Since you will not be using siding, you will have roughly ¾ inch of space behind the brick veneer, which is acceptable.

PREPARING THE WALL FOR VENEERING

On new work the sheathing is covered with 15-pound roofing felt nailed in place. Door and window frames are selected for use with brick veneer; they must project farther from the sheathing than standard frames.

On old work, measure how far the door frame and window casing project beyond the siding. If you don't have at least 1½ inches of clearance you should take one of two steps: remove the siding, or extend the door and/or window frame by nailing 1-inch-thick strips of wood to their exterior surfaces. The clearance is necessary to permit the veneer to stand clear of the wall and at the same time to permit the frame to overlap the brick by ¾ inch or so. A smaller overlap makes a positive seal at this point difficult to secure.

The next step is to trim the window and door sills so as to make their sides flush with the window casement and door frame.

Then you have to make certain the veneer wall will come up behind the fascia. To do this quickly and accurately, drop a plumb line from the interior side of the fascia to the masonry support. Mark the point of the bob on the masonry. Place a brick there and see if you have sufficient clearance behind it. If not, move the fascia away from the building the required distance, or cover the joint between the bottom edge of the fascia and the veneer with wood trim. This can be done only if there is an overhanging soffit to protect the trim from rain (see page 421).

LAYING THE BRICK

Start by dropping a plumb bob from behind the fascia at both ends of the veneer wall. Mark the front edge of the wall-to-be carefully. Then lay the first course dry to see how wide your joints should be to make the bricks come out exactly — if possible. Measure upwards and compute the precise horizontal joint width necessary to have the courses end up exactly beneath the soffit's edge. Install a story pole at one end of the wall. This will help you hold the courses to the correct height as you build.

The accepted method of terminating the veneer beneath a window sill is to finish with bricks laid on edge at an angle. Should the course immediately below the angle course come out incorrectly, cut the bricks lengthwise as necessary. The angled bricks will hide the cut edges.

In addition to laying the bricks very carefully so as to secure neat and accurate results, take care not to let any mortar protrude behind the bricks. You don't want the veneer to touch the wall. To reduce the possibility of mortar getting behind the bricks, skimp toward the rear side of the brick when you butter. Then use a stick to get behind the bricks and clean them as you go.

The veneer must be tied to the wall behind it with metal ties made for the pur-

A mason checks his guideline before starting a new course of brick veneer (left). Note the vertical line, dropped from the fascia, which serves as a guide at each end of the wall. Above, mason lays a closure brick to complete the course.

To aid water runoff beneath window sill, bricks are laid in tilted rowlock course.

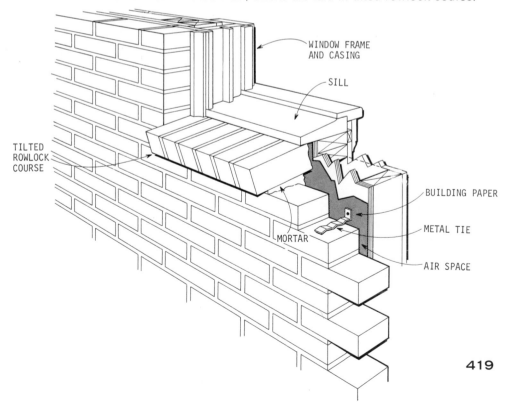

WINDOW FRAME AND CASING

SILL

TILTED ROWLOCK COURSE

BUILDING PAPER

METAL TIE

MORTAR

AIR SPACE

419

pose. Fasten the ties to the studs with eight penny common nails, as shown. Space the ties 32 inches apart horizontally and 16 inches apart vertically. Offset the rows of ties so that no tie is much more than 16 inches from any other tie.

Veneer corners are made in the same way as the corners of a single-brick wall. Just overlap the bricks as they meet at the corner. But since you are possibly going up 8 or 10 feet, take great care to hold the corner true. You will find a plumb line dropped from the soffit overhead will be of great help in doing this.

When you are only going up halfway on an exterior wall, you must flash the top of the veneer. For best results bring the flashing out over the top edge of the bricks and then down for about ½ inch.

When the veneer has to go over a doorway or window, it must be carried by a suitable angle-iron lintel.

Weep holes. In order to prevent moisture from collecting behind the veneer wall, there must be free air passage behind the wall. If air passage is going to be obstructed, it is advisable to provide weep holes along the bottom of the veneer. These

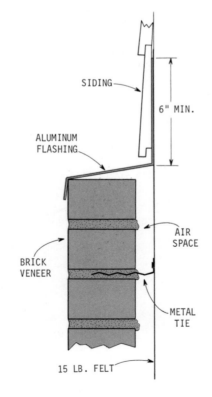

Method of flashing a partial veneer wall.

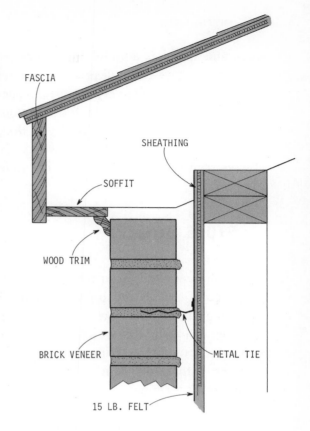

FASCIA

SHEATHING

SOFFIT

WOOD TRIM

BRICK VENEER

METAL TIE

15 LB. FELT

Properly positioned brick veneer will terminate behind the fascia (above). If the fascia
doesn't hide the top of the veneer wall, the
joint can be concealed with wood trim (right).

holes can be made by laying pieces of oiled rope between every fourth brick and then pulling out the rope as soon as the mortar stiffens a little.

INDOOR PROCEDURES

If you are going to lay the veneer directly on the same footing that is supporting the masonry wall you are going to cover, you do not require a space between the veneer and the existing wall. But you do have to bond to it. Therefore, clean the wall and the footing carefully and apply a bonding agent to both. Use mortar between the veneer bricks and the masonry wall to bond the two together.

If you do include a space you will introduce a little insulation and soundproofing between the two walls. However, you will have to use ties, which can be trouble-some, to fasten the veneer to the old wall. If possible, attach the ties to the wall with steel-cut masonry nails. If that is too difficult use a gun. These guns shoot nails into concrete. They must be used with extreme care as they can shoot a nail with as much force and speed as a bullet and as dangerously. *Do not attempt to shoot a nail into solid stone.* The gun will recoil dangerously.

If you are going to lay the brick veneer wall on a concrete slab which is not integral with the standing wall's footing, you must space the veneer away from the wall and tie it down as described. If you don't the veneer may crack.

Indoors you do not have to worry about the height of the veneer wall as you can always trim the top with wood. The sides, too, can be hidden by wood trim.

STONE VENEER

Stone veneer is used for the same reasons as brick veneer: to improve the appearance of a wall and to protect it from the elements, vermin, and fire.

Like a brick veneer wall a stone veneer wall can be placed in front of any type of existing wall — whether of wood, stucco, clapboard, or concrete block. The problems you will encounter working with stone veneer will be the same as those you will meet working with brick. The veneer must be substantially supported. The top of the veneer must terminate below the soffit, if it is outdoors. So much for the similarities. Now for the differences.

Unlike brick, the stone is usually placed close to or against the supporting wall. The face of the veneer is generally positioned about 6 inches from the backing wall. Since the stone is irregular, it is difficult to use a line as a guide. Therefore the supporting wall becomes the guide, and each stone is positioned by measuring from its face to the wall. Since the face of the stone is rarely smooth it is necessary to judge exactly how far from the wall each stone should be placed in order to produce an acceptably smooth surface. A great deal of judgment is necessary here, and it is advisable to go off to one side of the job to see how a temporarily positioned stone looks before you make that position final.

Since stone is almost nonabsorbent, you use a very dry mortar. Use 1 part regular cement to 3 to 4 parts sand. Mix thoroughly, but add so little water that the mix looks dry. However, there must be sufficient water to enable you to form a ball that will hold its shape.

Joints can be finished any way you wish. The joint between the edge of the stone veneer and whatever wood or metal siding it may contact is made close but is never sealed. There is no need for sealing since there is no space behind the stone veneer.

PART V

COMBINATION PROJECTS: Steps, Porches, Chimneys, and Fireplaces

The design and construction of steps, porches, chimneys, and fireplaces have been left to last for several reasons. Their construction involves a number of different techniques, all of which have already been covered. These masonry portions of a building are usually the last to be constructed, and since they are the most intricate and complex of possibly all masonry projects they rightfully belong at the end of the book. Their order of presentation is also related to their complexity. Chimneys are more difficult to construct than steps and fireplaces more difficult than chimneys.

In addition, fireplaces can be considered as a sort of test. If you can build one you will know that you have learned just about all this book has to offer on the subject of masonry.

MASONRY STEPS

There are any number of ways to construct masonry steps. Whichever way you select, two design requirements must be satisfied if the steps are to serve their purpose. One is the matter of human comfort, which limits the rise of each step to 8 inches and requires a tread depth minimum of 9 inches, but preferably 12 inches outdoors. The other is support. The steps must be properly supported or they will break up with frost and time. The exception, of course, are dry-masonry steps, described in Chapter 5.

To determine individual step rise, establish the total rise as suggested in Chapter 5. Then divide this distance experimentally until you find the number of steps necessary that will hold each riser to less than 8 inches. To determine individual tread depth, divide the horizontal distance between the start and finish of the steps by the number of steps you plan to build. Obviously, all this will take a little juggling of the figures.

When you have determined rise and tread you can select the type of staircase you wish to construct from the various designs suggested below, bearing in mind that you will probably have to vary tread and rise a number of times more before you are ready to begin construction.

These concrete-block steps have pre-cast concrete treads.

CONCRETE-BLOCK STEPS

When all you need is two or three steps the easiest and most economical way of constructing a masonry staircase is usually by building the steps of concrete block. When there is no frost, you can dig down past the topsoil and lay your block directly on the earth. If the soil is soft you can place 4-inch solid blocks beneath the cored block to spread the load. It isn't necessary to have a solid footing. When frost is a problem, you can simply excavate to the necessary depth and pour concrete in the hole to a depth of about 8 inches. Then build up on this base. There is no need for a form. The footing does not need to be more than a few inches larger than the block staircase.

Select and arrange the block so that you secure the rise and tread necessary. Do not worry about cutting and patching, since this will not be seen. Use 12-inch block

Two steps of concrete block after cement plaster has been applied. Soil will come up to the bottom edge of the still-wet plaster.

for 12-inch treads, and 14-inch block for 14-inch treads. Lay up the block with mortar as usual. Fill the cores with rubble. Then plaster the entire assembly with a ¾-inch-thick layer of mortar cement. Don't forget when you mortar the treads to give them a forward pitch, to keep them from collecting water.

Finish the treads with a float or even a brush, as explained in Chapter 14, just as if they were small slabs. Then round the nose, or edge, of the step with an edging tool. Leave the forms in place for at least forty-eight hours. When you remove the form, there may be some roughness on the risers. Fill in any holes with mortar and grind down bumps with a grinding tool.

For a better appearance you can lay down flagstone treads. Make the treads longer and wider than the supporting block by an inch all around. This will produce an overhang that makes the final job more attractive. The flagstone is held in place with mortar.

As an alternative to flagstone treads you can use precast concrete treads. You can purchase them in a number of sizes or make them yourself using a simple form and a high-sand mix $(1 : 2^{3}/_{4} : 3)$.

Should you want to reduce the number of blocks, you can make the treads self-supporting by incorporating reinforcing bars within each step when you cast it. If you support each step along its front edge all you need is one or two bars near the rear edge of the casting.

BLOCK AND CONCRETE STEPS

When the staircase is so large that a prohibitive number of blocks are required, consider combining block and concrete. This may be accomplished by pouring a footing of minimal thickness for each side of the staircase. The footing must go below the frost line but need not be much more than 8 inches thick and a little wider than the block. The sides of the staircase are then constructed of block laid on top of the footing. Most of the space between the block walls is filled with soil and rubble. Then a form is constructed using the blocks as guides and supports. One arrangement is shown in the accompanying illustration. Then the form is poured, using the high-sand mix suggested, starting with the bottom step. When just this portion of the form has been filled, it is tamped and floated and permitted to stiffen up for twenty minutes or so. Then the second step is poured, and so on. The wait between pouring steps is necessary, otherwise the wet concrete would simply overflow the lowest step form. Be careful not to overfill any step section, for if you do, you will have an identation at the rear of the step.

Concrete blocks can be used as supports for step forms, which are then partially filled with sand and rubble, and poured with concrete.

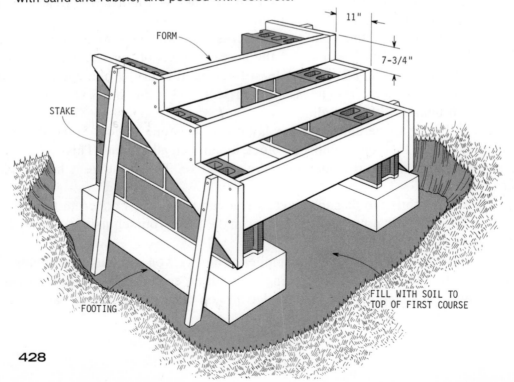

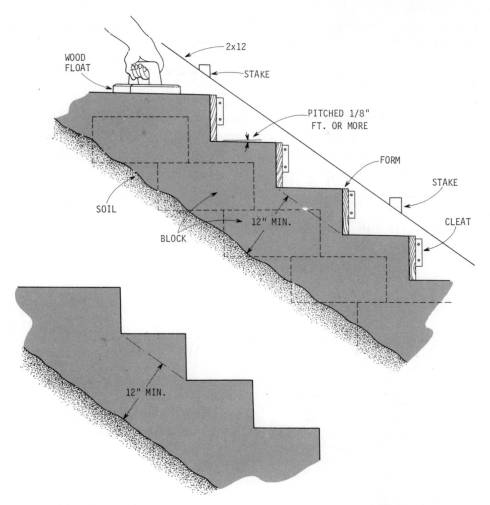

To complete the block-and-concrete steps, each section is poured and floated in turn, starting from the bottom and working up.

ALL-CONCRETE STEPS

The form generally used to construct steps of poured concrete is shown in the accompanying illustration. If the rise of each step is equal to the width of a standard board, it is convenient to build the form of boards. When step rise does not match a standard board width, it is usually easier to make the form of plywood.

429

When pouring only two or three steps and frost is not a problem, you can pour directly on firm subsoil. Excavate an area larger than your form; install and brace the form; make certain it is vertical and square, tight against the building or porch; and that the finished surfaces of the treads will pitch downward for proper water runoff. Generally a pitch of ¼ inch to the foot is sufficient.

Where frost is a problem, you have to excavate to a depth below the frost line. In such cases, pouring more than two or three solid concrete steps becomes costly. One solution is to fill the center of the form with rubblestone, making certain that the surface of the rubble comes no closer than 1 foot to the sides of the form. In this way you will be certain the completed steps will consist of concrete 1 foot thick on two sides.

Bear in mind that the surface of the inside of the form will determine the surface

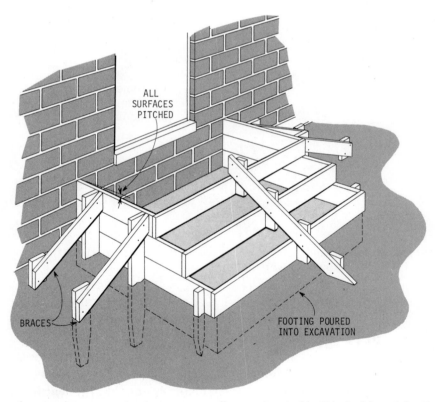

ALL SURFACES PITCHED

BRACES

FOOTING POURED INTO EXCAVATION

Forms for pouring concrete steps are usually constructed in this fashion. A footing is first poured into the excavation, and then the forms are poured from the bottom up. When the concrete has set up, the forms are removed.

Here sidewall forms are ¾-inch plywood panels. Note that both the sidewall and riser forms are well braced. Broken concrete will be used as fill before the concrete is poured. *Courtesy Portland Cement Assoc.*

A completed set of poured steps. The concrete has been floated and the edges finished with an edging tool.

of the completed steps. An indentation on the inside of the form will result in a bump on the surface of the steps. So, unless you are planning to parge (plaster) the concrete after you remove the form, make certain all interior form surfaces are smooth. To ease form removal, either wet the inner surfaces of the boards or coat them with oil before pouring.

To save money, you can use a 1:3:5 mix for that portion of the job that will be out of sight. The balance of the form should be filled with a 1:2¾:3 mixture.

Be sure to provide sufficient bracing for the form and to pour very slowly. And as suggested above, pour the bottom step and the area behind it first. Then wait for the bottom step to stiffen up before pouring the following step, and so on. If the sides of the form are high, poke a stick into the concrete as you pour to make certain the form is completely filled.

Finish the steps with a float or brush and an edging tool as previously suggested.

BRICK STEPS

The standard procedure for constructing bricks steps is to build a base of block or concrete first and then veneer it with brick. Obviously, a good-looking job necessitates a great deal of care in laying the brick. Not so obvious, but far more important, is the absolute necessity of planning the support very carefully. If you don't, you will end up with a patched brick job that won't do credit to the mason responsible.

Planning the support takes a little time because you have to work out the tread and riser requirements so that you do not end up with a partial brick. The support is stepped along its sides to provide a shoulder that carries the brick. Note that the brick treads never overlap the risers the way flagstone treads often do.

CUT-STONE STEPS

Steps made of cut stone are constructed exactly like brick steps in that the stone is supported by concrete blocks or poured concrete. The basic difference between the two is that flagstone is used for the treads and cut stone for the risers. Bear in mind when ordering stones that each stone must be laid in mortar, and that each mortar joint should be about ½ inch thick. Each stone's joining surface must be perfectly clean. Use mortar made from 1 part cement and 2 to 3 parts sand. Lime is not added to mortar that is used with stone.

BUILDING CUT-STONE VENEERED STEPS

1 The first flagstone tread is in place, along with the cut-stone risers for the second step. The mason is now spreading mortar to carry the second flagstone tread.

2 The mortar level is checked. Note the concrete blocks that form the steps and support the stone.

3 The second flag is laid in place. Here you can see the cutstone risers that have already been installed on the first and second steps.

4 A cut stone that is to be part of the following riser is positioned. After additional stones are laid alongside, the space behind will be filled with stone rubble and topped with mortar.

The steps of this stairway have been completed. All that now remains to be done is to veneer the sidewalls with cut stone. The spaces left at both ends of the flagstone treads will accommodate the sidewall stone veneer. Note the metal ties mortared into the block.

434

STOOPS
AND
PORCHES

A preliminary definition of terms may be useful here. A stoop is a raised entrance platform that projects beyond the wall of the building. A porch is a similar entrance platform, but it is protected by a roof of some kind. A veranda is simply another name for a porch. Most building departments are quite unconcerned about how and where you construct a stoop. But if you cover the stoop with a roof it becomes a porch and an integral portion of the building in the eyes of most building inspectors. Therefore, if your home's front wall is exactly on the legal setback line from the road, your stoop can project across that line, but your porch cannot. This restriction applies to side yards as well as to setback.

Stoops and porches serve a variety of useful purposes, some of which may come immediately to mind. A stoop provides a place to stand when the front yard is knee-deep in water or mud. Stoops make it easier to open and enter a front door; you don't have to reach up and step up as you would from a walk. And stoops, of course, can provide the base for an entryway, assuming it is acceptable to the authorities.

A porch is even more useful. It is a fair-weather extension of your home's living space—a place to sit on a summer's evening, a place for potted plants, an additional room, when enclosed, where you can eat your meals.

From a mason's point of view, the stoop and the porch are identical, except for size. Both consist of self-supporting (reinforced) concrete slabs supported by perim-

This small porch of poured concrete provides a pleasant entryway, a protected area for loading and unloading a car during a heavy rain, and a place to sit on warm summer nights. *Courtesy Portland Cement Assoc.*

eter footings. The method of construction depends mainly on whether the house is under construction or has been in existence for a number of years.

NEW CONSTRUCTION

Stoop foundations. When the top of the stoop isn't much more than 5 by 6 feet, the usual method of construction consists of supporting the building side of the stoop on corbeled concrete blocks (blocks that project out from a wall). These blocks should be below the frost line.

Assuming that the frost line in your area proves to be 4 feet or six blocks below the top of the foundation wall, construct a normal concrete-block foundation wall (see Chapter 24). When you have worked your way up to just six courses short of the top of the foundation, position two blocks within the foundation wall at a right angle to the wall. They should be so spaced that they will be in exact vertical alignment with the blocks comprising the sides of the stoop. At least half of these two blocks will stick out like huge sore thumbs, which is fine. Lay two more courses

436

of block and then corbel two more blocks exactly above the first two. The remaining blocks in the foundation wall are laid down as usual.

The next step consists of backfilling the foundation — replacing the soil that was removed — to a height about equal to the bottoms of the two lower corbeled blocks. Then a footing trench is dug beginning at one corbeled block, extending directly outwards from the building, then making a right-angle turn followed by another to return to the second lower corbeled block. A form is constructed within this trench for a footing that will be 10 inches wide and 8 deep. The top of the poured footing will be exactly level with the tops of the two lower corbeled blocks.

When the footing has hardened sufficiently, you can lay 6-inch block on it to

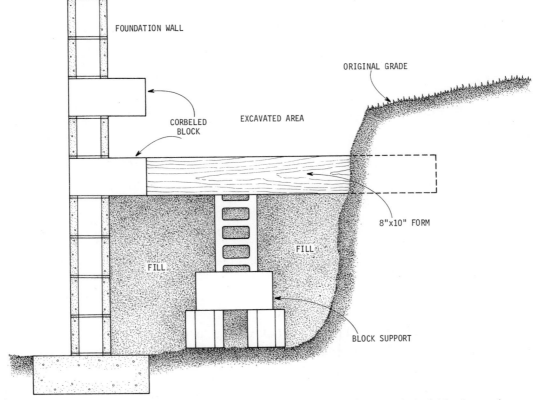

The side of the stoop or small porch can be supported by corbeled blocks and a poured footing (form is in place) extending from the corbeled blocks. When the span from the foundation wall to the original grade is more than 5 feet, position concrete block as shown before backfilling.

437

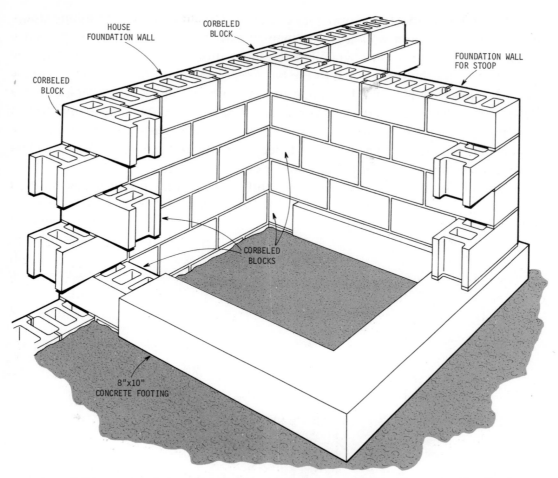

CORBELED
BLOCK

HOUSE
FOUNDATION WALL

CORBELED
BLOCK

FOUNDATION WALL
FOR STOOP

CORBELED
BLOCKS

8"x10"
CONCRETE FOOTING

A partially completed stoop foundation wall. Note the position of the poured footing in relation to the corbeled blocks.

build the sides of your stoop. The first block you lay is positioned on top of either of the lower corbeled blocks and next to the foundation wall. When you have laid two courses of stoop foundation blocks, the third course will again begin on top of a corbeled foundation wall block.

The stoop's block foundation should terminate about 6 inches below the desired height of the stoop. This allows room for the concrete slab that will be poured on the foundation. If you are going to flag the concrete, terminate the stoop's foundation 2 inches lower. More details about the slab will be given later.

Porch foundations and footings. A porch starts out simply as an extra-long stoop. Porch construction on a new house is begun at the same time as work on the house footing, since the porch foundation also rests on this footing. Start by making the house footing under the porch section of the foundation wall 6 inches wider than the rest of the footing. Then, as you lay the concrete block for the house foundation, lay adjoining and parallel courses of 6-inch block, locking every second course of 6-inch block to the house block with a girder or expanded sheet metal reinforcement. The girder or expanded metal is simply laid horizontally across the pairs of blocks as you go. This locks the two walls together.

When you have reached the frost line—bearing in mind you measure down from ground level to determine this, not the top of the foundation wall—you need to make a slight change in the 6-inch block wall you are building. Up until this level, your wall is as long as the total length of the porch. At this point, you reduce its length by 20 inches or a little more—10 inches on each side. Now continue up with the 6-inch block, leaving a shoulder or step on each side. Stop when you have reached a point 6 inches below the desired surface of the 6-inch slab you are going to pour.

The next step consists of digging a trench for a footing 10 inches wide and 8

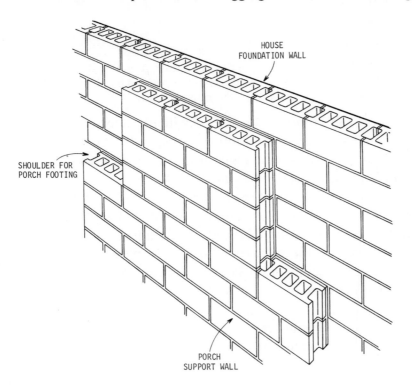

HOUSE
FOUNDATION WALL

SHOULDER FOR
PORCH FOOTING

PORCH
SUPPORT WALL

Porch foundation wall is built adjoining and parallel to the house foundation. Shoulders are left in the porch foundation to carry the footing.

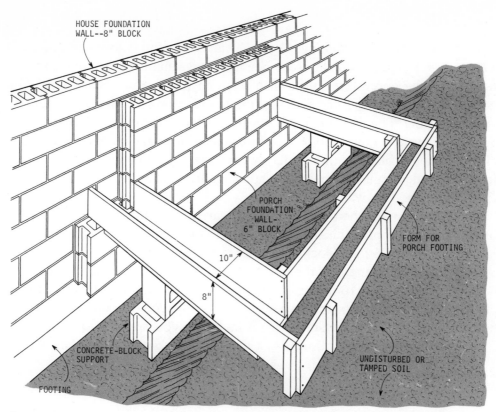

HOUSE FOUNDATION
WALL--8" BLOCK

PORCH
FOUNDATION
WALL-
6" BLOCK

FORM FOR
PORCH FOOTING

10"

8"

CONCRETE-BLOCK
SUPPORT

UNDISTURBED OR
TAMPED SOIL

FOOTING

The form for the porch footing is made of stock lumber to produce a footing 8 inches high and 10 inches wide. The form is supported by the shoulders of the block wall, by blocks positioned in the trench, and by the soil on the opposite side.

inches deep. Make the bottom of this footing trench level with the steps or shoulders on the 6-inch porch wall foundation. Start the footing at one step. Have it run directly outward from the building, make a right-angle turn, run the desired length of the porch and make a turn back to the second foundation step.

Next you have to reinforce that portion of the footing which will span the space between the foundation wall and the earth alongside the footing but separated by a gully. Do this by placing four half-inch-thick steel rods in each of the footing forms where they touch the foundation wall. Place the rods close to the bottom of the form, with their ends touching the foundation wall. The rods should be long enough to stretch a foot or more beyond the edge of the gully. Now you can pour the footing using a 1 : 2¾ : 3 mix.

When the footing has hardened sufficiently and you have stripped the form, erect the rest of the 6-inch block perimeter foundation that is to support the porch slab.

440

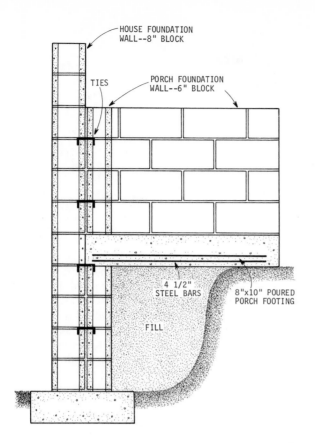

HOUSE FOUNDATION
WALL--8" BLOCK

TIES

PORCH FOUNDATION
WALL--6" BLOCK

4 1/2"
STEEL BARS

8"x10" POURED
PORCH FOOTING

FILL

Side view of a porch foundation wall on its footing. In this instance, the foundation wall has been built of 6-inch block. Four steel bars reinforce the footing.

Supporting the slab. In the case of a stoop you have two parallel concrete-block foundation or supporting walls separated by a hole. No support is offered by the building foundation wall and the fourth side of the stoop is open. This is where the steps will go. Steps are discussed in Chapter 30. In the case of a porch, you have four supporting walls surrounding a hole. The problem in both cases is how to support the wet concrete of the slab long enough for it to harden.

The usual method of providing this support is to fill the hole with soil to the proper height, which is level with the top of the supporting foundation walls. Some masons tamp and level the soil at this point. Others just level the soil within the walls and cover it with a layer of tar paper. When there isn't too much soil on hand, you can provide support by building a flat form of 1-inch lumber nailed to 2x4s spaced no more than 2 feet apart. The width and length of the form are 1 inch less than the space between the supporting walls. The form is then placed on upended concrete blocks in such a way that the 2x4s rest on the blocks, which should be no more than 2 feet or so apart.

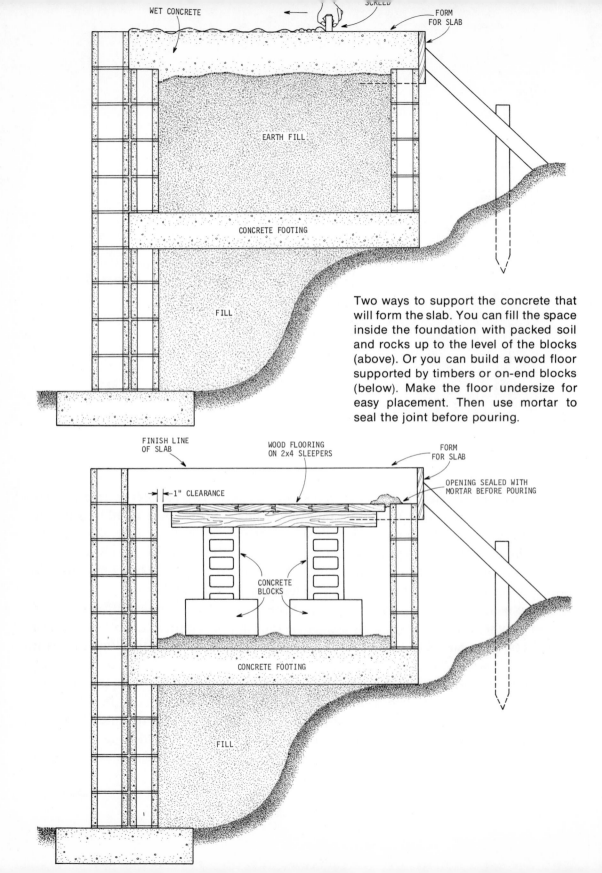

WET CONCRETE

SCREED

FORM FOR SLAB

EARTH FILL

CONCRETE FOOTING

FILL

Two ways to support the concrete that will form the slab. You can fill the space inside the foundation with packed soil and rocks up to the level of the blocks (above). Or you can build a wood floor supported by timbers or on-end blocks (below). Make the floor undersize for easy placement. Then use mortar to seal the joint before pouring.

FINISH LINE OF SLAB

WOOD FLOORING ON 2x4 SLEEPERS

FORM FOR SLAB

OPENING SEALED WITH MORTAR BEFORE POURING

1" CLEARANCE

CONCRETE BLOCKS

CONCRETE FOOTING

FILL

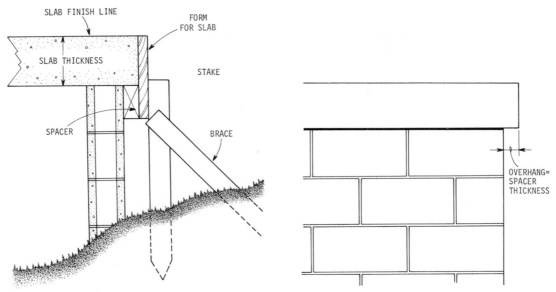

To make the slab overhang the foundation, use a spacer between the foundation and the form board.

Another method of supporting the slab while it is being poured and cured consists of filling the space with rubble masonry topped by a layer of crushed stone.

Whatever method you use, bear in mind that the support you construct must be made level with the top of the porch or stoop foundation blocks, and that whatever material you use will remain hidden. Remember also that this temporary support does not eliminate the need for steel reinforcements within the slab.

After you have provided a support for the slab to be poured, a form is constructed for it. Use 1x8 tongue and groove boards with the grooved edge on top. The form is positioned around the foundation and pressed tightly against the concrete blocks with stakes. Make the top of the form 6 or more inches above the top edges of the foundation wall blocks. The distance above the block determines the thickness of the slab. Pitch the sides of the form to provide rain water runoff. A pitch of about $\frac{1}{8}$ or $\frac{1}{4}$ inch to the foot is sufficient.

Reinforcing the slab. At this point you are ready to mix and pour the slab, except for the steel work, which cannot be omitted or skimped. The reason is that this slab differs from the others previously discussed — walks, patios, driveways — in that it is supported only by its edges. It must therefore contain sufficient steel to prevent it cracking when the supporting material settles and sinks because it is fill—soft, uncompacted earth—or else rests on fill.

The accompanying tables give the required quantity of steel for various slab widths and thicknesses. Some masons bend the ends of the steel bars and hook them into the holes in the blocks for presumably greater strength. Some do not. In any event, the steel bars should be positioned about 1½ inches from the bottom of the slab and wired together to prevent them from shifting during the pouring of the concrete.

SIZE AND SPACING OF MAIN BARS AND CROSSBARS FOR A SLAB SUPPORTED ON TWO SIDES

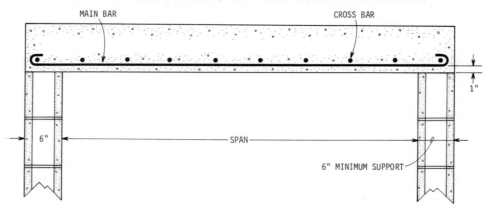

ONE-WAY REINFORCEMENT ONLY.
(Must use deformed bars; bars made for reinforcing concrete)

Slab thickness	Bar	Span	Bar spacing	Cross bar spacing	Cross bar size
4"	3/8"	4'	10"	18"	3/8"
4	3/8	5	8	18	3/8
4	3/8	6	6	18	3/8
4	1/2	8	7	12	3/8
4	1/2	10	4	6	3/8
6	3/8	4	12	18	3/8
6	3/8	5	10	18	3/8
6	1/2	6	10	14	3/8
6	1/2	8	7	12	3/8
6	1/2	10	5	10	3/8

Bar sizes and spacing based on a "residential" load, meaning 50 pounds per square foot or no more than a dozen fat people to any one porch or stoop.

SIZE AND SPACING OF BARS FOR A SLAB SUPPORTED ON FOUR SIDES

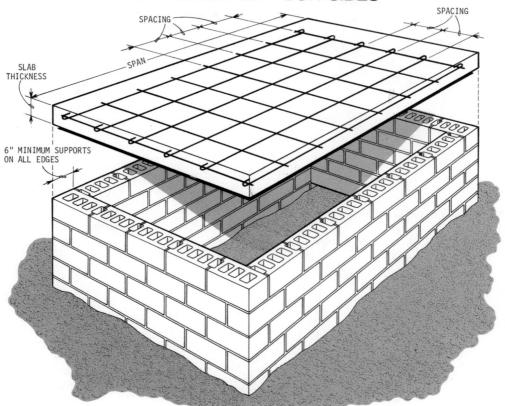

TWO-WAY REINFORCEMENT

If the slab is not square, use longest dimension in calculations.

Slab thickness	Bar size	Span	Spacing between bars
4″	$3/8$″	5′	18″
4	$3/8$	6	14
4	$3/8$	8	10
4	$1/2$	10	10
6	$3/8$	5	20
6	$3/8$	6	18
6	$3/8$	8	14
6	$1/2$	10	12

Bar sizes and spacing base on a "residential" load, meaning 50 pounds per square foot.

Typical deformed steel bars made for reinforcing concrete.

BAR SIZES

(Inches)	Numbers	Weight (lbs. per. ft.)	Cross-sectional area (sq. in.)
1/4	2	0.166	0.05
3/8	3	0.376	0.1105
1/2	4	0.668	0.1963
5/8	5	1.043	0.3068
3/4	6	1.502	0.4418
7/8	7	2.044	0.6013
1	8	2.670	0.7854
1/square	9	3.400	1.0000
1 1/8 square	10	4.303	1.2656
1 1/4 square	11	5.313	1.5625

Pouring the slab. Use a 1 : 2¼ : 3 mix a little on the stiff side. Pour very slowly and gently. Rap the side of the form after all the concrete is in place to help it settle. Screed, tamp and float as usual. If you are going to flag the concrete, quit after screeding.

If you are planning to install an iron railing, gently poke a length of wood 1½ inches square about 6 inches deep into the spot where you want the rail post to go. This should be at least 4 inches in from the edges of the slab. Do this after screeding and tamping. Let the wood remain until the concrete sets up, then gently turn and pull it out. Next, go over the surface near the hole with the float again to make it even. Do not let the wood remain in the concrete more than a couple of hours. If it stays in until the concrete is very hard you will have to chisel it out. If it rains, the wood may swell and crack the concrete.

OLD CONSTRUCTION

Stoops. When a building has been in place twenty years or more it is reasonably certain that the fill around the house foundation has settled sufficiently to support a stoop. On this assumption, dig a footing trench around the perimeter of the projected stoop. Pour an 8-inch-thick, 10-inch-wide footing and construct your stoop from 6-inch concrete block following the procedures already suggested.

If the stoop is to be no more than a foot or so high and you don't have to go down very far to get below the reach of frost, construct a form for all three sides of

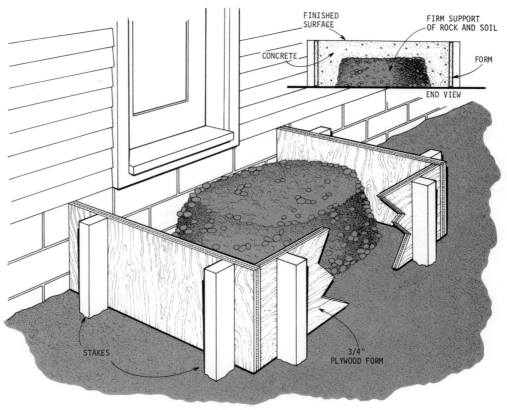

When the soil is firm and there is little frost, you can pour a slab for a stoop against the wall of the house. Build a plywood form as shown, partially fill it with rubblestone and soil, and then pour the concrete.

the stoop (the building's foundation wall forms the fourth side). Let the bottom edges of the form rest inside the trench you dig. Fill the interior of the form section with rubblestones and soil. Then slowly pour concrete between the wood form and the inner mound of rubble and soil. Remember the warning in Chapter 19 about what happens to forms that are filled too quickly. When you have filled the form to within about 6 inches of the top, hold off for an hour or so to let the concrete set up a little. Then fill the rest of the form, completely covering the mound of rubble and soil until the concrete is flush with the top of the form. Now you can screed across the entire form, tamp and float finish.

Should you need a step, you can make one from a slab of concrete. Build a form as long as necessary, 2 feet wide and 8 inches high. Fill it with concrete, and when the concrete has hardened, remove the form and push the concrete step in position alongside the stoop.

Porches. Since porches are generally much larger than stoops, you will be taking somewhat of a chance if you depend on the old fill alongside the house alone for support. Definitely the best and safest, if the hardest, method is to excavate to a little below the existing house footing and then to pour a new, additional footing alongside the old one. The foundation for the porch on the building side is then erected on the new footing. As you lay the new 6-inch block, knock a few holes into the old wall, say one every 3 feet every second course of block, and tie the new wall to the old by hooking bent pieces of strap across the top of the new block and into the holes in the old block. Then mortar the holes closed.

The rest of the porch may be constructed exactly as previously suggested for porch construction on a new building.

CHIMNEYS
AND
FIREPLACES

CHIMNEYS

Chimneys built inside a building work a little better than those outside the building, especially in winter, since the warmer the air within the chimney the greater its draft. Moreover, inside chimneys yield their heat to the building and not to the outside air, which of course results in fuel savings. However, in the summer, you have the heat of the chimney to contend with, and of course inside chimneys occupy valuable floor space.

Another point of difference is the difficulty and cost of construction. Inside chimneys are hidden; you can hack and patch as you wish. Outside chimneys are exposed; you have to work carefully or at least plaster your work.

Whether you construct a chimney according to an architect's plan or design your own, it must satisfy the following requirements if it is to work properly, be safe and pass building-department inspection.

Chimneys must rest on solid concrete or rock. A one-story chimney — a chimney up to 20 feet in total height — should rest on a 10-inch-thick base. A two-story chimney — a chimney 35 feet in height — should rest on a base at least 12 inches thick. The 10-inch-thick base should extend at least 4 inches on each side of the bottom of the chimney and the 12-inch base at least 6 inches on each side. No matter what the base dimensions of the chimney may be, no base should be less than 4 feet square, unless you are on solid rock.

To determine whether or not your base is sufficiently large, compute the weight of the chimney at 125 pounds per cubic foot for brick masonry, 60 pounds per cubic foot for block masonry, and 140 pounds per cubic foot for poured concrete. Soft clay will safely support one ton per square foot. For additional data on the load-bearing capabilities of various soils see Chapter 16.

The house frame, wood floors, walls, and roof must be constructed to provide a full 2 inches of clearance all around the chimney. (Some fire codes require greater clearance.) The space can be filled with rock wool or similar incombustible insulation.

The chimney must be lined with vitrified clay at least 1 inch thick. This is a fire precaution. All fires produce soot, which will ignite unpredictably and burn so fiercely and at such a high temperature that the resultant gases cannot escape through the chimney opening fast enough to relieve the high pressure. This pressure can force flames through cracks in masonry. The clay liner reduces this possibility, and the clay, being smooth, collects far less soot than masonry.

The top of the chimney must be a minimum of 3 to 4 feet above the top of the roof. The exact minimum figure depends on the position of the chimney top in relation to the roof top. Some codes may call for different minimums.

While the flue lining's cross section may be round, square, or rectangular, it must be large enough to pass the volume of combustion gases produced by the furnace or fire. The accompanying table suggests minimum flue cross-sectional areas for various applications. Larger flue sizes always ensure a better draft.

Each flue within a chimney structure must be completely separate from all other flues. There must not be any interconnection between them. When possible, the flues should be separated by masonry. When this is not possible, flue joints should be staggered by 7 or more inches to minimize the possibility of gases from one flue entering another.

The chimney itself must be perfectly vertical. The contained flues are best when vertical but can be offset if necessary. Make the offset as gentle as possible, since sharp turns reduce draft. If there are two flues within the chimney and one has to be offset, the fireplace flue should be offset rather than the gas- or oil-furnace flue.

Construction. Chimneys can be constructed from brick, hollow or solid concrete block, or a combination of all three, with or without cut-stone trimming. You can

A typical concrete-block interior chimney with a single flue designed to be connected to an oil-burning furnace.

Partially completed interior chimney. Note how space between liner and block is filled with mortar.

451

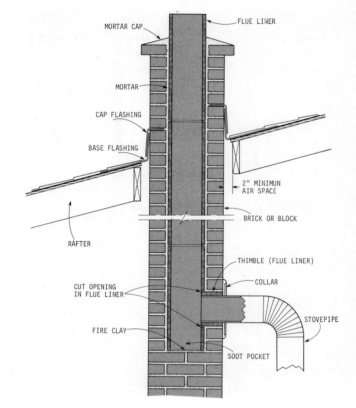

Cross-section of a typical chimney installation.

also use chimney block, which is special block made for the purpose with a suitable hole for the flue liner in its center. To make a chimney from these blocks you merely lay them up one on top of the other. No matter what material you select, chimney construction remains basically the same.

If your chimney's base is integral with the foundation wall, build both at the same time, but do not build the chimney portion higher than the foundation. Instead, wait until the house is framed and sheathed. Then you can use the building as a support for your scaffold and ladder and as a guide for your masonry work. If your chimney stands free inside the building, pour its footing at the same time as you pour the other footings. Then wait until the building is up before commencing construction.

Start by laying out the four corners of the chimney on the poured footing. If you have constructed the base of the chimney along with the foundation wall, this is the time to double-check before building higher. At this point you can still shift your chimney an inch or two if you need to.

To make certain the corner marks are accurate and to help you hold the chimney vertical as you go up, drop one or more plumb-bob lines from the building itself.

	Rectangular Outside diameter	Circular	Bellmouth Inside diameter
Gas-fired central heating systems	8x8 inches	8 inches	8 inches
Stoves, ranges, space heaters, hot-water heaters	8x8	8	8
Coal- or oil-fired central-heating furnace	12x12 .	12	— —

Suggested minimum flue lining sizes for various applications.

The accompanying illustrations show how blocks and bricks may be used to lay up a chimney. Just take care to make the joints solid and true. Don't stint on mortar quantity or quality. To cut flue lining, stand it on end and pack it tightly with wet sand. Then place a mason's chisel on the desired line of cut and keep rapping the chisel with a small sledge hammer as you slide the chisel along. Proceed just as you would if you were cutting flagstone. When the flue liner has been grooved all the way around it will break.

Exterior chimneys are separated from the house sheathing by a layer of building paper, then the chimney is constructed tightly against the paper. If the chimney is of brick or cut stone these materials are not used on the side of the chimney that is against the building.

The last flue liner is cut so that it will project 6 or more inches above the top of the chimney. Then it is installed cut end down. One way to finish the chimney is to give its top a steeply pitched mortar cap several inches thick.

Fireplace Width	Height	Depth	Square Flue Lining	Round Lining
28	28	16	12 × 12	10
30	30	16	12 × 12	10
34	30	18	12 × 12	12
41	30	18	12 × 16	
47	36	18	16 × 16	

Suggested minimum flue lining sizes for fireplaces of various openings and depths. All dimensions are in inches.

CONSTRUCTING A FIREPLACE AND CHIMNEY

1 The fireplace footing has been poured at the same time the building footing was poured. Here, the blocks that will form the fireplace foundation are laid in place.

2 The blocks are filled with rubble masonry before laying the bricks of the fireplace platform. This is going to be a basement fireplace.

3 The fire bricks are laid on the platform, which will be the inner hearth of the fireplace. The mason is tooling the mortar joints.

4 A double-walled steel fireplace liner is positioned over the inner hearth. The foundation wall is then constructed around the liner, as shown, with insulation at the corners.

5 Constructing the chimney. This design calls for a single-brick wall which surrounds the flue liner. The space between will be filled with rubble mortar. Note that the building has already been sheathed.

6 Rubble and mortar is dumped into the space between the flue liner and the exterior brick wall. The mortar mix generally used consists of 1 part cement, 3 parts sand and just enough water to enable the mortar to be formed into a ball. Mortar must be added very carefully so as not to move the recently laid bricks.

7 Positioning a section of flue liner. Liners are joined with a thin mortar joint. When the chimney bricks are laid above this liner, the next section is installed.

8 The brick wall of the chimney goes up. The single flue liner is now surrounded by mortar; the rubble is covered.

9 As chimney height increases, masons use a scaffold to keep mortar and bricks handy.

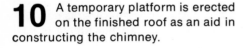

10 A temporary platform is erected on the finished roof as an aid in constructing the chimney.

FIREPLACES

Like the chimney attached to it, a fireplace has to be constructed according to known principles of combustion and building-code safety regulations.

The inner hearth, the surface on which the fire will rest, must be made of fire brick laid up in fire clay or a similar fire-resistant mortar. The walls of the inner hearth must be made of either fire brick or metal. The entire hearth must be solidly supported by a vertical masonry column, the center of which may be hollow for the collection of ash. The entire fireplace, as well as its chimney, must stand free and

458

clear of all combustible material by at least 2 inches or more, if the local code so specifies.

The rear wall of the inner hearth must tilt forward so that a shelf at least 5 inches wide is formed above the hearth. For best results the center of the inner hearth should be directly under the center of the flue. There must be an adjustable damper, its front edge flush with the opening leading to the smoke chamber. The size of the flue must be sufficient for the fireplace opening. If the flue is too small, the draft will be poor and the chimney won't pull.

In addition there should be an outer hearth at least 16 inches wide and extending a minimum of 8 inches beyond the sides of the fireplace opening. The outer hearth can be made of flagged concrete cast integral with the base of the inner hearth or it can be of masonry supported by suitable framing.

A brick chimney on a poured concrete foundation. Bricks have been set in mortar that has been darkened with added color. Marks left by foundation form will be hidden by cement plaster.

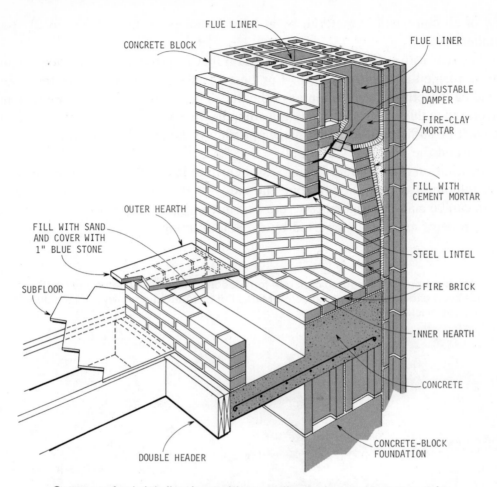

FLUE LINER

CONCRETE BLOCK

FLUE LINER

ADJUSTABLE DAMPER

FIRE-CLAY MORTAR

FILL WITH CEMENT MORTAR

OUTER HEARTH

FILL WITH SAND AND COVER WITH 1" BLUE STONE

STEEL LINTEL

SUBFLOOR

FIRE BRICK

INNER HEARTH

CONCRETE

DOUBLE HEADER

CONCRETE-BLOCK FOUNDATION

Cutaway of a brick fireplace with a cantilevered concrete support for the hearth. The perimeter of the outer hearth is built up of bricks, the interior filled with sand and covered with bluestone.

Constructing a brick fireplace with a raised hearth. The fireplace is built between vertical 2x4s which serve as guides for the bricks and as nailers for the Sheetrock that will cover the walls. The bricks at lower left will form a portion of the raised hearth. These bricks are laid directly on the poured concrete basement floor.

Construction. Use masonry or poured concrete to construct the base of the fireplace and its associated chimney. The base can be solid or it can have a hollow center through which the ashes are dumped and collected at the bottom.

The fireplace is constructed on top of this platform. First the fire-brick hearth is laid down. Then the fire-brick side walls are constructed. To make the curve for the rear wall, draw a guide line on the side bricks using a bent stick. Next the outer walls of the fireplace may be laid up and the space between the fire brick and the outer walls filled with mortar and masonry rubble. The damper is positioned and locked in place with mortar. Then, if the design calls for it, a steel lintel is installed to carry the front section of the chimney above the fireplace. Next the smoke box is constructed and plastered smooth with a layer of fire clay. Then the rest of the chimney can be erected.

If you are using a metal liner, place it above the fire-brick hearth and build the chimney around and above it. The liner is more expensive than the fire brick, but much easier to work with. The double-walled liners allow the passage of air behind the fire and so extract a greal deal more heat from the fire than it would otherwise radiate into the room.

The inner face of the fireplace can be veneered in brick or cut stone. A lintel is used to carry the masonry across the fireplace opening. Any mantelpiece you may choose to install rests on the brick veneer.

461

APPENDIX

REPAIRING CONCRETE AND MASONRY

FOUNDATION CRACKS

Foundations crack because the earth beneath their footings settles unevenly. The cracks themselves are easily filled, and if the earth stops settling the cracks will remain closed.

Clean the cracks with a small cold chisel and hammer. Wet the concrete — block or poured — thoroughly and wait until there is no free water remaining in the cracks. Then mix 1 part mortar cement with 3 parts sand and add sufficient water to make a mix with the consistency of ice cream. With a trowel force the mortar into the cracks. Then, with the edge of the trowel, smooth the patch flush with the wall.

If the earth continues to settle unevenly, the cracks will open again. There are only two cures: to hire a contractor to pump fresh concrete at high pressure beneath the foundation footing, or to increase the footing area by placing concrete beneath it by hand.

To increase footing area, dig a trench along one side of the building which extends several feet below the footing. Then remove a small section of the soil beneath the footing and replace it with concrete. Repeat the operation on each side until the entire footing rests on a concrete base. In this way you increase the area of soil on which the foundation and its footing rest, and by going deeper perhaps reach virgin soil that will not settle.

OPEN JOINTS IN BLOCK AND BRICK WALLS

Open joints differ from cracks in that each opening is confined to the height of the brick or concrete block. The open joints are caused by careless masonry work. Either the mortar was much too dry when applied, or the brick or block was moved a little before the mortar hardened. The resultant loss of strength is usually not important, especially in a block wall. However, open joints are easily closed. This is how it may be done.

Wet the wall thoroughly. Mix 1 part mortar cement with 2 parts sand and add sufficient water to produce a mix with the consistency of ice cream. Wait until there is no water on the surface of the wall nor in the openings, then force the mortar into the open spaces. You will probably find a small pointing trowel most convenient for the job, though you can secure the same result with a small stick. To keep bricks clean, cover them with masking tape.

SANDY CONCRETE SURFACES

If a concrete walk or drive was permitted to dry out before it had set, or too much sand was used, or the mix wasn't mixed thoroughly, or the slab was troweled too much, the surface is often very weak and, as a result, sand is released.

One cure consists of brushing the concrete clean of all loose sand, wetting it thoroughly, and then covering it with a soupy mixture of cement and water. Then float or trowel the surface. (see Chapter 14 for details on finishing concrete). The aim here is to force the liquid cement into the concrete so as to bind the top grains of sand more firmly to the slab. When you have finished using the trowel, the surface of the slab should not be higher than it was before.

Another cure consists of brushing the slab clean and then applying a commercial cement sealing compound to the surface.

A third cure consists of brushing the slab and simply painting the offending surface.

The trouble with these three cures is that they presuppose the flaws that produce the sandy surface to be limited to the topmost layer of concrete. When this is not the case, they do not produce permanent results.

ROUGH SURFACES

A rough concrete surface can be made smooth by covering it with a layer of mortar. First the old surface is carefully cleaned. If the surface is in the process of spalling and flakes of concrete are coming off, a small sledgehammer and a cold chisel are used to remove all loosely adhering concrete. If a portion of the slab is smooth, you must either roughen that area with a hammer and cold chisel to produce a pockmarked surface or coat the entire slab with a bonding agent before applying the mortar.

The second step consists of flooding the slab with water and waiting until the concrete has soaked up all the water and none remains on its surface. Generally, this is when the bonding agent is applied.

While waiting for the water to disappear, mortar is prepared by mixing 1 part cement with 3 parts sand and adding sufficient water to secure a soft ice-cream consistency.

A ¼-inch-thick layer of mortar is spread over the slab with a trowel. The mortar is pressed firmly against the slab and finished as described in Chapter 14. Since you will be working on a relatively flat and smooth surface, you will find it easy to make the surface of the mortar you apply flat and smooth.

An alternate method, advisable when there are large, uneven areas in the surface of the old slab, consists of erecting a form around the slab and using the form as a guide to produce a new, higher but flat and smooth surface.

The form boards are pressed against the sides of the old slab with stakes driven into the ground. The top of the form is positioned about ½ inch above the top of the slab. Then the mortar mix is applied to the surface. A straight-edged board is used to screed the surface of the fresh mortar level with the top edges of the form. The new surface is finished as usual.

HOLLOWS OR LOW SPOTS

If a concrete surface develops a hollow in one spot, the slab can be repaired without replacing the existing concrete.

Clean the low spot and roughen with the help of a hammer and chisel, chipping

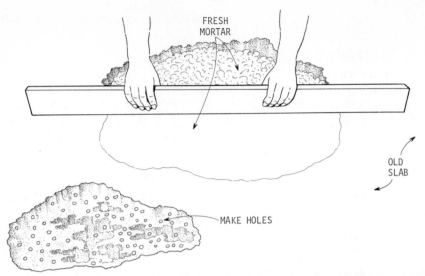

FRESH
MORTAR

OLD
SLAB

MAKE HOLES

To fill a shallow hollow in a concrete slab, clean the depression and roughen it by chiseling small holes in the concrete. Wet it down and fill it with mortar. Then screed it level with the balance of the slab as shown.

shallow pock marks every inch or so. As an alternative to chisel surgery, you can use a bonding agent.

In either case, flood the slab with lots of water and wait until all the water has been absorbed. Make a batch of mortar consisting of 1 part cement, 3 parts sand, and sufficient water to bring the mix to the consistency of ice cream. With a shovel and trowel spread the mortar over the lowest areas of the hollow, but do not bring the mortar to slab level.

Next, mix a second batch of mortar using the same proportions, but first run the sand through a sieve of window screening and strain out the large grains.

With the first layer of mortar in place and still fairly soft, cover the balance of the hollow with the second batch of mortar. (If the hollow is shallow, you can omit the first layer of mortar and just use the screened-sand mortar.) Level the mortar with the slab by screeding it with a straight-edged board, then finish as usual.

BADLY BROKEN CONCRETE

Very often a bad winter will break up a section of a driveway or walk, or a heavy rain will wash away the supporting soil and a passing auto will crack the concrete badly. To repair, you do not have to remove all the old concrete. You can rebuild the slab.

With a sledgehammer break the sunken or unsupported section of the slab into 1-foot pieces. The trick is to hit the concrete repeatedly in the same spot until it breaks. If you scatter your blows you will be at it all day. (Wear safety glasses, as the chips will fly.)

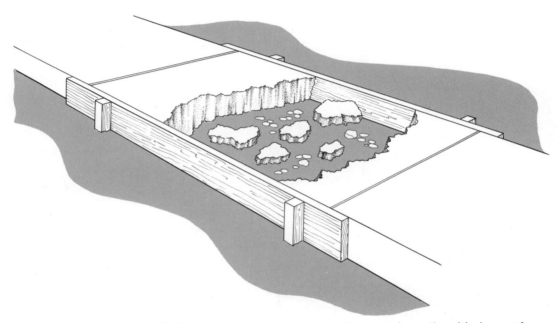

When the slab is badly broken, construct a form as shown and use the old pieces of concrete as fill. Then proceed just as you would when making an ordinary slab walk.

Next, treat the broken slab area exactly as if it were a section of a new walk or driveway you are going to build. The only difference is that instead of using crushed stone or gravel, you use the large pieces of broken concrete.

Install one or two boards, as necessary, on edge against the sides of the unbroken slab, holding them tightly in place with stakes. The boards should be flush with the unbroken surface of the slab. Soak the pieces of old concrete and the edges of the unbroken slab thoroughly, then fill the form with a 1 : 2¾ : 3 mix, which will be easier to work with. Screed and finish as previously described.

SUNKEN SLABS

When a concrete walk sinks into the ground you can raise it to its original level by following this procedure.

If the slab has not cracked along its crack-control grooves, you must crack it yourself. This can be done by striking the concrete along the groove with a small sledgehammer and cold chisel.

Then lift the sunken sections of the walk with the help of a pickax or a crowbar, and slip some bricks or stones underneath to hold the sections of the slab in position. Force gravel or crushed stones beneath the raised sections of concrete, filling the space.

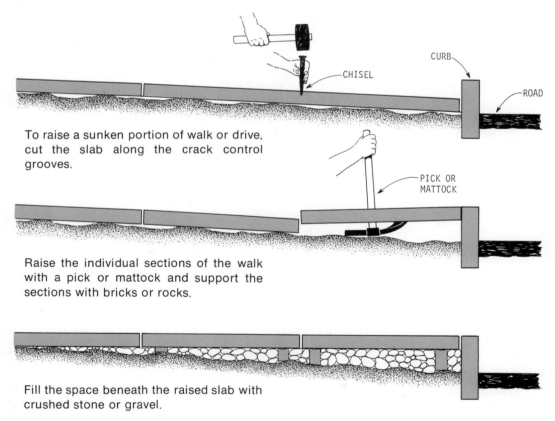

To raise a sunken portion of walk or drive, cut the slab along the crack control grooves.

Raise the individual sections of the walk with a pick or mattock and support the sections with bricks or rocks.

Fill the space beneath the raised slab with crushed stone or gravel.

The once sunken walk is now raised to its proper level and fully supported on a bed of small stones. The cracked grooves will not be noticed.

LOOSE AND BROKEN FLAGS IN MORTAR

When water seeps beneath flags set in mortar and freezes, several things may happen: the mortar in the joints may crack and be forced out of the joints; the flags may be loosened; the flags may be forced out of the mortar; the flags may be broken, even shattered.

Open joints. With a narrow cold chisel and small sledgehammer, remove all the cracked or loose mortar from between the flags. Take great care not to loosen any flags. With a stiff brush clean the joints of all particles of mortar and soil. Then soak the joint openings with water and when the water has disappeared, fill and finish the joints with mortar as described in Chapter 15.

Loose flags. Remove all the old mortar surrounding the loose flag. Then remove the flag. With a cold chisel and small sledgehammer chip away all the mortar that was beneath the loose flag. Chip until you are down to the surface of the supporting concrete slab. Brush the opening clean and soak it with water. When the water has disappeared, proceed to re-lay the flag following the same procedure described in Chapter 15. Give the mortar a few days to harden, then fill and finish the joints.

Broken flags. It is best not to attempt to pry up broken flags as there is a strong possibility that you will damage or loosen adjoining flags. First, carefully remove the mortar surrounding the broken flag, then remove the flag. Replace it with a new flag following the same procedure as above.

LOOSE AND BROKEN CONCRETE BLOCKS AND BRICKS

If the block or brick (b/b) is in the middle of a wall and it is loose, meaning you can move it a little, this is proof positive that it carries no load and for all practical purposes can be ignored.

If for some reason you merely want to immobilize a b/b, force a piece of solder wire between it and the b/b next to it. The lead is easily forced into the opening and will not deteriorate or absorb water.

On the other hand, if you want to close whatever openings there may be between the loose b/b and the wall, fill the opening(s) with mortar made from 1 part mortar cement, 2 parts sand, and sufficient water to produce a consistency equal to that of ice cream.

Broken or crumbling b/bs in the middle of a wall can also be ignored if you wish. They carry no load or the pieces could not be moved. To replace them, they must be removed ever so gently so as not to loosen the rest of the wall. This can be done with a narrow-blade cold chisel and a hammer. Chip slowly and gently away at the mortar until the unwanted brick or block is freed. Then clean the entire space of old mortar. At least remove sufficient old mortar to permit you to fit the new wall unit into place with space for some fresh mortar.

Mix up some mortar as just described but add a little more water to secure a consistency equal to soft ice cream. Apply a layer of mortar to the bottom and sides of the opening in the wall. Apply a layer of mortar to the top of the b/b. Then gently force it into place. Next, finish the joint with a small pointing trowel or a jointing tool.

Loose b/bs on top of a wall should not be ignored as they may fall and injure someone.

The replacement procedure and the mortar used is the same as just described. However, if adjoining b/bs are shaky or a little loose it is advisable to remove, clean, and replace them with fresh mortar. For strong results remove all the old mortar. Do not lay fresh mortar on top of old.

If only a single b/b is loose and you don't want to bother to mix mortar, remove just enough of the old mortar to permit the b/b to fit. Then use an epoxy formulated specifically for masonry to hold it in place.

RESURFACING MORTAR JOINTS (POINTING)

All masonry joints exposed to rain and frost crumble away. Joints between flags deteriorate with each passing year because cracks caused by one winter's freeze make it easier for water to enter the following year.

Joints in a vertical structure break down at a fairly constant rate. The surface of the mortar crumbles into sand and the wind blows the sand away, and as this happens the mortar recedes farther and farther in between the blocks, bricks, or stones.

Unless the mortar is defective, the loss of joint strength in proportional to the loss of mortar. Since most walls are about 12 inches thick and are capable of carrying many times their load, the loss of a ¼ or even ½ inch from the surface of the mortar joint does not measurably weaken the wall. Thus there is no need to point a wall unless the joint has lost more than a ½ inch or so of mortar.

To point, start by removing the loose, crumbly layer of sand that may be present on the surface of the mortar joints. This is easily done with a narrow-blade cold chisel and a hammer. Then, with a stiff brush, clean the joints thoroughly.

If the weather has been dry for the past few days, hose the wall and wait an hour for the water to be absorbed.

Mix 1 part mortar cement with 2 parts sand and add sufficient water to secure a consistency a bit stiffer than that of ice cream. (You want a slightly dry mix to reduce the chance of staining the face of the wall.) With a hawk (or an inverted steel finishing trowel) to hold a quantity of the mortar, and either a small pointing trowel or a caulking trowel, force mortar into the joints. Finish the joints with the same trowel or with a jointing tool.

If in the course of chipping away at the old mortar to uncover solid mortar you find the old mortar to be sandlike all the way through, you will have to take the wall apart and rebuild it—at least the weak part. Pointing—adding fresh mortar over the old, weak mortar—will not strengthen the wall. More than likely pointing will weaken the wall since the very act of forcing fresh mortar between the bricks or stones tends to force them apart.

You will encounter this condition most frequently in the topmost courses of a brick chimney. When you do, take the chimney apart, starting with the top bricks and work your way downward until you can no longer lift bricks out of place. *Force* the next course of bricks loose and rebuild the chimney, as described in Chapter 32, from that point up.

LOOSE RAILING POSTS

If the concrete surrounding the bottom end of the post is unbroken, the post can be tightened in its hole by any one of several methods:

1. You can pour molten lead into the space between the post and the walls of the hole.

2. You can force lead into the space between the post and the walls of the hole. Use either pieces of solid wire solder or lead wool, which can be purchased at a plumbing supply shop. Pound the lead into the spaces with a thin rod and a hammer.

3. You can fill the spaces with grout made by mixing 1 part cement with 2 parts sand and enough water to secure a consistency equal to thick soup.

4. You can fill the spaces with epoxy mixed with sand or use the special epoxy made to patch concrete.

When the concrete surrounding a railing post breaks off, construct a form as shown and pour fresh concrete or mortar to rebuild the broken area.

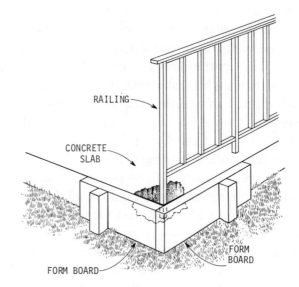

Broken concrete. Railing posts are usually positioned a few inches from the edge of the supporting concrete slab. When the concrete breaks, a large chip usually falls away and leaves one side of the post hole open.

To repair, remove all the loose pieces of concrete and dust. Let the post remain in the hole, but force it away from the side of the hole with a small piece of stone. Cover the break in the side of the concrete slab with a board, its top edge flush with the surface. Hold it in place with stakes.

Wet the hole and broken concrete edges and wait until the water has been absorbed. Force a 1:3 mixture of cement, sand, and water into the space between the slab, post, and board. Do not let anyone or anything touch the post and rail for three or four days.

BROKEN CONCRETE CORNERS AND ENDS

The ends and corners of concrete walks and driveways break off because of insufficient support. Trying to rejoin the piece to the rest of the slab with mortar is a waste of time. The piece will soon break off again. The correct repair consists of pouring new concrete on top of a good base.

Remove the broken piece of concrete. Then excavate a small area around the end of the slab to a depth of about 6 inches, also cleaning out some of the soil beneath the slab. Fill the excavation with new gravel.

With a star drill and a small sledgehammer or an electric drill and a masonry bit, drill ½-inch holes, 6 inches deep, into the edge of the old slab. If the slab is a driveway, drill the holes about 1 foot apart; if a sidewalk, about 2 feet apart. Don't drill the end holes too close to the sides of the slab. Thus, the broken end of a 10-foot-wide driveway would require 9 holes; a 4-foot-wide walk would require 3 holes.

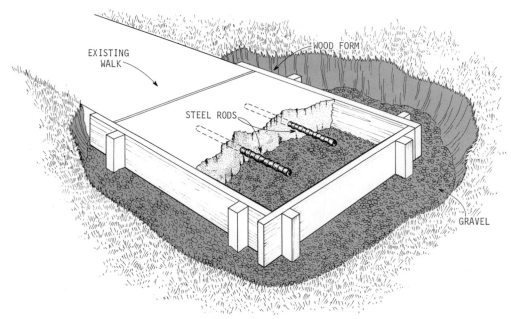

When rebuilding the end of a walk or drive, excavate the soil around the end of the slab and fill with gravel. Install reinforcing rods. Then build a form of the appropriate size to pour the end of the slab.

Purchase as many lengths of ¼-inch steel reinforcing rods as there are holes. How far the rods should protrude from the edge depends on the amount of new concrete you need to pour to rebuild the slab. The rod ends should not be less than 5 inches from the edge of the new concrete. Force water into the holes. Then force grout — 1 part cement, 2 parts sand, and enough water to make a thick soup — into the holes. Then push and hammer the bars into the holes.

Construct a form the same size as the missing end or corner of the slab (see drawing). Thoroughly wet the broken edges of the slab. Fill the form with concrete, screed, and finish.

Wire-reinforced concrete. When wire-reinforced concrete breaks, the wire holds the pieces together. But since the break was caused by the soil beneath the slab settling or washing away, the broken piece tilts downwards.

One way you can try to correct this fault with a minimum of labor and expense is to dig beneath a portion of the tilting concrete and force a few stones underneath to lift the piece of concrete back into position. Then remove the soil under the balance of the broken end and force more stones into place.

BROKEN STEP AND PORCH CORNERS

If no more than the tip of a corner breaks off and you can find the piece, cement it back into place with epoxy cement. Just make certain the materials to be joined are dry and keep the joint warm until the epoxy sets.

If the portion broken off is large, so large it will be stepped on during the normal

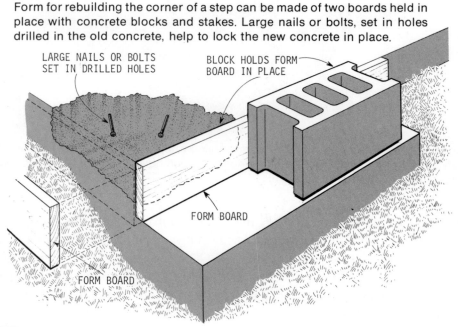

Form for rebuilding the corner of a step can be made of two boards held in place with concrete blocks and stakes. Large nails or bolts, set in holes drilled in the old concrete, help to lock the new concrete in place.

LARGE NAILS OR BOLTS
SET IN DRILLED HOLES

BLOCK HOLDS FORM
BOARD IN PLACE

FORM BOARD

FORM BOARD

course of foot traffic, it is useless to cement it back into place. A new corner must be cast from concrete and tied to the balance of the structure with steel.

This is the way it can be done. Drill one or more ¼-inch holes 1 inch deep into the surface of the break, using either a star drill and hammer or an electric drill and masonry bit. Construct a wood form as illustrated. Insert heavy nails or bolts into the holes, long enough to extend at least 1 inch beyond the broken surface of the concrete. If the corner you are going to cast is very large, use longer bolts.

Next, wet the broken surface of the concrete. Mix 1 part cement with 3 parts sand and enough water to make a mixture with the consistency of ice cream. Pour the mix into the form. Screed the top of the mixture level and finish as usual.

DEFECTIVE CAPPING

Tile. Cracked tile capping can be repaired by sealing the cracks with either epoxy cement or a little grout made by mixing 1 part cement with 1 part sand and lots of water to get a soupy mix. Wipe the excess off before the cement hardens. Loose tile can be replaced and held in place with mortar made by mixing 1 part mortar cement with 2 parts sand and enough water to make a mixture with the consistency of ice cream.

Concrete. Missing sections of concrete capping can be replaced by simply wetting the area down and molding a new cap in place. Use 1 part cement, two parts sand, with just enough water to make it plastic. Apply with any kind of trowel or float. You can even use a flat board if you have nothing else on hand.

Cracked capping on the tops of chimneys, high walls, and other invisible places can be sealed with a thin layer of asphalt. Just be careful not to apply so much that it runs down when heated by the sun.

EFFLORESCENCE

The white powder that sometimes forms on bricks and concrete is called efflorescence. It is harmless and you can ignore it if you want to. To remove it, try hosing the surface down. If water alone won't do it, try scrubbing the wall or floor with a steel brush and then hosing it down. If this still doesn't remove the white salts, mix 1 part muriatic acid with 20 parts of water (always adding the acid to the water). Put on rubber gloves and carefully apply the acid. When the acid has done its job, remove it with lots of water.

FASTENING TO CONCRETE AND MASONRY

There are a number of ways of permanently fastening things to concrete and masonry. Some of the more frequently used methods and hardware are described in this section.

Fiber plugs. These short tubes are used for holding wood screws in concrete, brick, and block walls. The tube is inserted in a hole drilled in the wall. When the screw is driven into the center of the tube, the tube expands, locking itself and the screw permanently in place.

Select a plug to match the size of the wood or sheet-metal screw you plan to use. Then, using a small star drill and a hammer, or an electric drill and a masonry bit, make a hole in the wall deep enough to accept the length of the screw to be inserted, having a diameter just equal to or a small fraction of an inch larger than that of the plug. Push the plug into the hole, then thread the screw into the hole in the center of the plug. If the plug turns as you turn the screw, remove the plug, Hold it in your hand and thread the screw partway into the plug. Thus expanded, you can now return the plug to the wall and thread the screw as far as you wish into the plug. The friction between the inner wall of the hole and the plug will keep it from turning. Screws held by fiber plugs can be removed and replaced as desired.

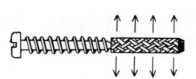

When a screw is threaded into a fiber plug, the plug expands and develops pressure against the walls of the hole.

Fiber Plug

If the plug turns in its hole when you try to thread the screw into it, remove the plug and enlarge it a little by partially threading the screw into it. Then replace the plug in the hole and finish threading the screw into the plug.

Lead and plastic shields and anchors. All of these are installed the same way as the fiber plugs and grip the wall by virtue of the screw or bolt expanding the shield or anchor as it is screwed home. These anchoring devices are available in a wide variety of sizes and designs. The plastic anchors usually are less expensive and are made for small screws and machine bolts. The lead shields and anchors, which usually have internal threads, are used for larger screws and machine bolts.

Lead Anchor

Plastic Anchor

Courtesy Rawlplug Co.

479

INSTALLING A LEAD OR PLASTIC ANCHOR

Drill a hole in the masonry and insert the anchor.

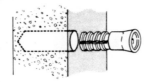

If necessary, strike the anchor lightly until end is flush with the surface of the wall.

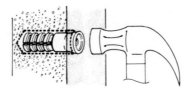

Place the fixture in position and insert the screw. Driving the screw into the anchor expands it and locks it in place. *Courtesy Chicago Expansion Bolt Corp.*

Expansion shields. Also called lag shields, these differ from the fastening devices described so far. They consist of a tube cut longitudinally and joined at one end. They are used for large screws and lag screws (sometimes called bolts) which have very coarse threads.

An expansion shield with a lag screw in place. Note how the entrance of the screw expands the two halves of of the shield. *Courtesy Chicago Expansion Bolt Corp.*

The installation procedure remains the same. First the shield is selected to fit the screw, then the drill is selected to fit the shield. The hole is drilled deep enough to accept the shield and bolt. The shield is inserted in the hole. When the lag screw is threaded into the shield, the shield expands and holds in the wall.

Expansion plugs. These differ from the anchors and shields just discussed in that the gripping action of the plug against the walls of the hole can only be produced by actually tightening the bolt against the object it is to hold. The plug is expanded when the bolt forces the bottom section of the plug into the upper section. Also, these fasteners are made to use only with bolts.

A standard expansion plug. When the bolt pulls the cone wedge (left) into the sleeve, the sleeve expands, locks the plug in the hole.

Double-action expansion plug. This design has two wedges which result in greater holding power. This plug is recommended for loads that are in line with the plug. *Courtesy Rawlplug Co.*

Expansion nuts. These are also called calk-ins, machine screw anchors, and expansion nut anchors. Whereas expansion plugs and shields can be used in bottomless holes, expansion nuts can only be used with holes that have bottoms. The nut is expanded permanently in the hole by pounding it with a pipe-shaped little tool called a plug set. The hole must have a bottom or the nut won't expand. The bolt can be removed when necessary. The plug remains in place.

An expansion nut that is locked in place by hammering.

INSTALLING AN EXPANSION NUT

A hole is drilled and the nut is inserted.

A setting tool is positioned in line with the plug. The setting tool is then struck a few times with a hammer. This expands the lead sleeve, locking it in the hole.

A machine bolt is threaded into the hole and the fixture fastened in place. *Courtesy Chicago Expansion Bolt Corp.*

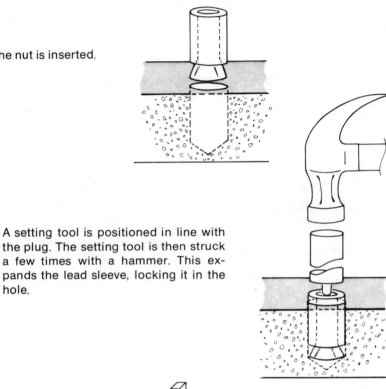
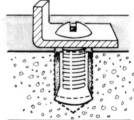

One-piece expansion bolts. This masonry fastener is very useful when a number of heavy-duty fasteners are required. You only have to drill a hole and drive the fastener into place. This fastener holds by virtue of the spring action of its split body. It can only be used in hard masonry and should not be used near the edge of a concrete slab or brick wall. Note also that it cannot be removed without considerable effort, so position your holes carefully.

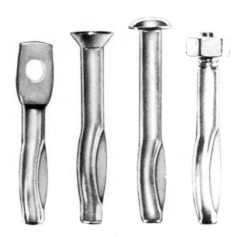

Four types of one-piece expansion bolts. *Courtesy Rawlplug Co.*

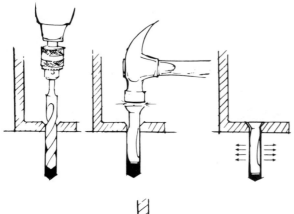

To install a one-piece expansion bolt, drill a hole thrugh the object to be fastened and into the masonry. The diameter of the hole should be the same as the diameter of the bolt. Drill the hole ¼ inch or deeper than the length of the bolt. Then insert the bolt and drive it home.

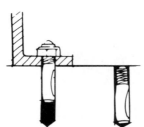

If you want to remove the object fastened to the masonry, use a stud bolt. Drill the hole so deep that, when you remove the object, you can drive the stud bolt completely into the hole. *Rawlplug Co.*

Select a drill of the same size as the shank of the fastener. Drill the hole at least ¼ inch deeper than the full depth the fastener will be driven. If you wish later to remove the object that is fastened, use the stud expansion bolt. Position the fixture, run the nut partway up on the stud. Drive the fastener into its hole. Then tighten the nut. If you plan to remove the fixture at some future date, make the hole so deep that you can drive the fastener below the surface.

Nails. Round, pointed steel nails are available for masonry work, but driving them into concrete and mortar is a difficult job. They cannot be driven into tile or stone.

The task can be made easier in a number of ways. You can hold the nail with a pair of pliers while you strike it, preferably with a small sledgehammer or a heavy carpenter's hammer. You can drive the nail through a board and into the concrete; then split the board away from the nail and drive it farther in, if you need to. You can use one of the new driving tools which hold the nail safely while you strike it. Or you can use a nailing gun. This is a special gun powered by cartridge blanks that shoots nails. Great care must be used since the gun can send a nail through the air with all the killing power of a bullet. For safe and proper operation the nose of the gun must be pressed against the board or whatever is to be nailed in place.

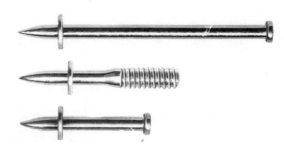

Three types of masonry nails presently available. *Courtesy Rawlplug Co.*

HOW TO USE A DRIVING TOOL

Insert the head of the nail into the tool. Position the point of the nail against the masonry.

Tap the striker, the central shaft of steel, until it touches the nail head. Then strike the pin with a hammer, driving nail into concrete. *Rawlplug Co.*

484

Precautions: Masonry nails are made of tempered steel. They do not bend, they break and fly every which way. Always wear protective glasses when driving masonry nails. Do not drive the steel nails into joints in a weak wall. The nails may crack the wall.

Drilling holes in masonry. Holes can be drilled in masonry in several ways. One is to use a star drill. Another is to use a masonry bit in an electric drill, and still another is to cut the hole with a diamond-point chisel.

When using a star drill, keep the drill at right angles to the surface of the work at all times. After every hammer blow, give the drill a partial turn in either direction. If you don't turn the drill as you strike it, eventually you will find that you cannot remove the drill from the hole. Star drills usually make a hole slightly greater than their largest diameter.

Three star drills. The larger drill is hand held and used for holes ⅜ inch and larger. The two smaller drills mount in a steel holder and are used for making smaller holes. *Courtesy Rawlplug Co.*

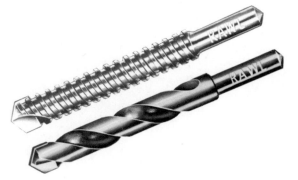

Masonry bits are tipped with tungsten carbide, which has a very high melting point. Nevertheless, if the tip gets too hot, the joint will soften and the tip will loosen. *Courtesy Rawlplug Co.*

Making holes in concrete with a masonry bit in an electric drill is simple enough. Just apply an even steady pressure and rest the drill when it becomes hot. To make holes 1 inch and more in diameter, which you may need for installing iron railing posts, use a hollow-center masonry bit.

Large-diameter holes can also be made with a chisel in a surprisingly short time — not much longer than it takes to drill them — with a diamond-point cold chisel and a small sledgehammer. Hold the chisel almost parallel to the surface of the masonry and remove small chips of stone at a time. The edges of the hole will not be perfectly smooth and vertical, but the hole will be useful, nonetheless.

Index